D0502703

MECHANICS' INSTITUTE
❦ MECHANICS' ❧
MERCANTILE LIBRARY

Courtesy of salon.com

A SHEARWATER BOOK

At the Ends of the Earth

At the Ends of the Earth

A History of the Polar Regions

Kieran Mulvaney

ISLAND PRESS / Shearwater Books

Washington • Covelo • London

A Shearwater Book
Published by Island Press

Copyright © 2001 by Island Press

Shearwater Books is a trademark of The Center for Resource Economics.

Library of Congress Cataloging-in-Publication Data

Mulvaney, Kieran.
   At the ends of the earth : a history of the polar regions / Kieran
Mulvaney.
         p. cm.
Includes bibliographical references (p.   ).
   ISBN 1-55963-908-3 (cloth : alk. paper)
   1. Polar regions—History. I. Title.
   G580 .M85 2001
   998—dc21

                                        2001001599

British Library Cataloguing-in-Publication Data available.

Printed on recycled, acid-free paper ♲

Manufactured in the United States of America
10 9 8 7 6 5 4 3 2 1

To Sallie,
Melanie, and Dan

Contents

Prologue

A little less than two and a half thousand years ago, a Greek ship set sail from the port of Massilia—present-day Marseille, in the south of France—on a journey of discovery that the Massilians hoped would lead to greater understanding of the distant lands and little-known peoples with which they traded. Under the command of Pytheas, a noted astronomer of the time, the ship sailed through what we now call the Strait of Gibraltar, but which the Greeks, with greater flair, referred to as the Pillars of Hercules; turned north along the Breton coast; and made landfall in Britain. It lingered there awhile, and during this time Pytheas explored the country's interior, reaching as far as Belerium, "Land's End," on the tip of Cornwall. Much of the island, he recorded, was swathed in forests and swamps, above which peeked hilltops where the inhabitants had carved gwents, or open clearings. The people raised wheat, corn, and barley, and they fermented grain to make a kind of wine; they mined tin, kept domestic animals, possessed metal weapons, and rode in wooden chariots adorned with bronze and gold.

Returning to the ship, Pytheas continued his voyage and reached the Orkney and Shetland Islands, far farther north than any Greek explorer before him had gone. But Pytheas was not satisfied. During his British sojourn, he had learned of a place called Thule, six days' sailing to the north, where the sun never set in summer nor rose in winter, and he meant to find it.

The northern realms had long excited the imagination of his people. Many Greeks believed the region was home to a people called the Hyperboreans, who lived "beyond the north wind." These Hyperboreans, noted the poet Pindar, writing roughly a century before Pytheas, held "feasts out of sheer joy. Illnesses cannot touch them, nor is death foreordained for this exalted race."

1

Unfortunately, Pytheas' own account of his journey is lost, bar a few fragments. What we know of his voyage comes largely from later writers—Poly-
bius and, following him, Strabo—and they treated most of his assertions with
scorn, disbelieving his descriptions of the places he visited. We thus do not
know exactly where Thule was. Some have suggested it lay on the Norwegian
coast; others, that it was nothing more exotic than the northernmost of the
Shetland Islands. The Irish monk Dicuil insisted, on the basis of writings of
eighth-century clerics, that Thule was Iceland, a view that still commands much
support.

We also cannot be sure that Pytheas reached Thule and, if not, how far he
did go. Indications are that he did not travel as far north as he wanted to, for he
reportedly found his way blocked by a thick barrier—although whether that
barrier was of ice or fog, we can but speculate. He did apparently observe that
"the seas about the region were of a strange substance, neither sea nor air, on
which it was not possible to walk or to sail," suggesting he had encountered sea
ice, the frozen surface of the polar ocean. He may even, as at least one historian has argued, have reached the pack ice off the eastern coast of Greenland.

Whatever Thule's true identity, and however close to Thule Pytheas came,
one thing seems clear: Pytheas was the first recorded European to reach the
fringes of the Arctic. It would be several centuries before any others even came
close.

✳ ✳ ✳

About a thousand years later and many thousands of miles away, a young Polynesian adventurer named Ui-te-Rangiora slipped his canoe into the water and
steered south. For weeks he traveled, navigating by the stars at night and perhaps following the cues of migrating birds by day. Legends of the time spoke of
an earlier explorer, Te Aratanga-nukū, who had glimpsed a southern land his
people would later settle and know as Aotearoa, "Land of the Long White
Cloud": a land more widely known today by the less satisfying name of New
Zealand.

Perhaps Ui-te-Rangiora set out in search of this Aotearoa; if so, there is no
indication that he succeeded. (In fact, Polynesian colonization of the country
probably did not begin until around the eighth or ninth century.) He is, how-

ever, said to have continued far to the south, into a gray, stormy ocean with mountainous, foam-tipped waves, and with ice strewn about its surface. He had no words to describe what he was seeing, as it was beyond the realm of his people's experience. Falling back on the closest comparison he could muster—a common Polynesian plant and the starchy white substance it yields—he pronounced the ocean to be *Tai-uka-a-pia:* the sea with foam like arrowroot.

Acclaimed Polynesian historian Sir Peter Buck has cast doubt on whether Ui-te-Rangiora went very far, suspecting instead that when Pacific voyagers "struck cold weather by getting too far south [they] would turn north again because of their scanty clothing." A more likely scenario, according to Buck, is that tales of a frozen sea far to the south reached Polynesia via later European explorers, and these were incorporated into existing legends and oral histories.

Perhaps, then, Ui-te-Rangiora's voyage is a product only of mythology and imagination. But given the Polynesian proclivity for traveling extraordinarily long distances and colonizing far-flung islands throughout the Pacific Ocean, it is not wholly improbable that he or one of his countrymen—perhaps even one of the early settlers of Aotearoa—was the first to dip a tentative toe into the frigid ocean around Antarctica.

✳ ✳ ✳

Whatever the veracity of the tales told about Pytheas and Ui-te-Rangiora, their existence speaks to the grip the polar regions have had, for many centuries and more, on the imagination of cultures far removed geographically from either the Arctic or the Antarctic. That same fascination has revealed itself throughout the ages in sometimes fanciful accounts of journeys to the ends of the Earth. The Irish monk Saint Brendan, for example, supposedly sailed far to the north during the sixth century in search of the "Promised Land" of the saints. The Norwegian king Harald Hardraada—later slain by the forces of England's King Harold at Stamford Bridge in 1066—is reported to have led a voyage of exploration north into the unknown ocean, withdrawing in time to escape "the vast pit of the abyss." And Icelandic annals from the year 1194 contain oblique reference to finding the "cold edge" or "cold coast." Slowly, over time, myth and reality fueled and blended with each other as legendary accounts of the polar regions inspired explorations in search of the truth, which in turn informed further fictional writings.

In 1719, the Englishman George Shelvocke was rounding the tip of South America when he was driven farther south by a storm. Although he did not actually reach Antarctica, he experienced enough "prodigious seas," "misty weather," and "islands of ice" to convince him that this was not a place in which he particularly wanted to tarry. During their time in the Southern Ocean, he and his crew "had not the sight of one fish of any kind . . . nor one sea-bird excepting a disconsolate black albatross, who accompanied us for several days, hovering about us as if he had lost himself." Shelvocke's second-in-command, obsessed with their traveling companion and convinced it was a bad omen, shot it, "not doubting," noted Shelvocke, "that we should have a fair wind after that."

Far from abating after the bird's death, the winds that had pushed the ship into this unpleasant stretch of ocean continued to beat mercilessly; eventually, however, ship and crew returned home safely. Although Shelvocke's voyage is little known today, the image of a possessed sailor killing an albatross lives on, having joined with later accounts of events in icy southern seas to inspire Samuel Taylor Coleridge's famous 1798 poem "The Rime of the Ancient Mariner."

A little less than two decades after Coleridge completed the "Ancient Mariner," just as Britain's Royal Navy was about to embark on a prolonged series of explorations in the Arctic, Mary Shelley set the opening and climactic scenes of her classic novel *Frankenstein* amid the ice of the north. The book's narrator, Robert Walton, travels in search of a polar paradise but instead sails his ship into what seems certain to be an icy tomb. The crew is on the verge of mutiny when, to their astonishment, a weak and emaciated Victor Frankenstein staggers out of the mist, later followed by the creature that has pursued him to the very ends of the Earth. At the book's conclusion, the monster stands triumphantly over the corpse of his dead creator, drifting toward the North Pole on an ice floe and vowing to build a flaming funeral pyre to consume them both.

The polar regions made cameo appearances in other nineteenth-century classics. In Charlotte Brontë's *Jane Eyre,* for example, the eponymous heroine becomes immersed in the descriptions of the Arctic in Thomas Bewick's *History of British Birds* (a celebrated ornithological work of the time), and finds her mind drifting off to contemplate "that reservoir of frost and snow, where firm fields of ice, the accumulation of centuries of winters, glazed in Alpine heights above heights." And they took center stage in still others. In 1835, Hans Christian Andersen used the Arctic in his fairy tale "The Snow Queen." In the 1860s, Jules Verne set two novels in the Arctic; thirty years later, he wrote one about the Antarctic, a sequel to an 1837 novella, *The Narrative of Arthur Gordon Pym,* by an equally esteemed writer, Edgar Allan Poe.

Alongside these fictional works stand factual accounts by explorers who sought to shine a light on the remote polar realms. Many such narratives contain greater drama than anything dreamed up by the most imaginative of authors—not surprisingly, for those who strove to explore and conquer the Arctic and Antarctic confronted powerful adversaries: regions where conditions are as hostile to human survival as anywhere else on Earth. For daring to undertake such challenges, many paid with their lives—none more famously than British explorer Robert Falcon Scott and four comrades. The British, demonstrating a seemingly inherent preference for heroic failure over outright success, have long gloried in the tale: not of the explorers' achievement of the South Pole in January 1912, for in that they were beaten by a team of Norwegians, but in their subsequent deaths—and, particularly, the death of Lawrence "Titus" Oates.

Weak, cold, and almost lame from frostbite and gangrene, Oates woke up on the morning of March 13, 1912, and, according to Scott, announced, "I am just going outside and may be some time." With that, he staggered out of the tent and into a howling blizzard. His body was never found. Oates' death was probably suicide, brought on simply by the fact that he could no longer bear the excruciating pain and had no desire to stretch out his inevitable demise. But the inference that most of the British public chose to take from Scott's description of events, and the interpretation that has lingered in the national psyche, was of a selfless man, having recognized that his condition was proving burdensome to his comrades, sacrificing himself stoically for the common good. The names of Oates and, indeed, of the entire expedition became bywords for much that was perceived to be worthy about Britain at the time of empire: battling earnestly against hopeless odds, striving to succeed in the face of certain failure, and, when forced to confront the inevitable, accepting it with dignity and pride and no hint of rancor. As Scott proclaimed in his final written message to the public, "this journey . . . has shown that Englishmen can endure hardships, help one another, and meet death with as great a fortitude as ever in the past."

But today, the British Empire has largely faded from favor, and the twenty-first century finds more to admire in another polar hero: Ernest Henry Shackleton, a contemporary of Scott's and, like him, a naval officer, but a man cut from very different cloth. Scott was a creature of the establishment, scion of a family with Royal Navy traditions and handpicked to lead Britain's Antarctic endeavors. Shackleton, an Anglo-Irishman and son of a doctor, faced opposition from Scott's backers, who believed that the glory of Antarctic discovery was Scott's destiny and his alone. Scott could be imperious and foul tempered, and although his men publicly—and in many cases genuinely—expressed admiration and affection for him, there were those among them who viewed him with contempt. Shackleton presented himself as more a man of the people, and he was a natural leader; among his comrades, he inspired almost universal devotion.

Shackleton, in other words, is the kind of hero we are more comfortable eulogizing in this day and age, and in recent years, his star has come to eclipse Scott's. His cause is aided by the fact that his most ambitious voyage culminated in perhaps the greatest voyage of survival ever described. His 1914–1917 British Imperial Trans-Antarctic Expedition, which sought to be the first to traverse the continent, saw its hopes of success dashed when the expedition ship, the *Endurance*, became trapped in the ice floes of the Weddell Sea and was crushed. For five months, the twenty-eight members of the crew drifted slowly north on an ice floe before setting out in the three whaleboats they had rescued from the ship and landing on remote Elephant Island. There, twenty-two of the men sheltered beneath two overturned boats while Shackleton and five others took off in the third boat for South Georgia Island. After navigating 800 miles of angry Antarctic waters, they reached their destination only to find that they were on the opposite side of the island from the whaling station where they hoped to find help. Shackleton, Frank Worsley, and Tom Crean then crossed mountains that other men thought could not be crossed, staggered into Grytviken, and organized a rescue party for the men on Elephant Island.

It was a remarkable example of courage and determination, and it is one that continues to inspire. In the late 1990s alone, Shackleton was the subject of several new books and a critically acclaimed museum exhibit that toured the United States; at the time of this writing (October 2000), it is rumored that his story will be given the Hollywood treatment. But if Shackleton has emerged as the first among equals, he has not done so entirely at the expense of his fellow explorers; his is the rising tide that lifts all boats. Polar explorers, and even more so the polar regions themselves, continue to appear—even, it seems, with ever greater frequency—in popular culture.

Andrea Barrett's best-selling 1998 novel *The Voyage of the Narwhal* superimposes fictional characters on the historical search for John Franklin, a British explorer whose Arctic expedition disappeared in 1845. Peter Høeg's *Smilla's Sense of Snow,* a murder story that centers on a detective who is part Danish and part Greenlandic, uses the Arctic as a context to examine relations between the two nations. More fantastically, John Carpenter's *The Thing*—a remake of an earlier movie that was itself an adaptation of the John Campbell short story "Who Goes There?"—portrays the terror unleashed in an Antarctic research station by a murdering, shape-shifting creature from another world. The denouement of the *X-Files* movie sees Fox Mulder rescue Dana Scully from an alien spaceship hidden beneath the Antarctic ice cap.

And so it continues. Two and a half millennia after the Hyperboreans and the voyage of Pytheas, the polar regions remain a source of intrigue and fascination. Yet much of that fascination reflects the fact that even now—even with the Arctic, the Antarctic, and their surrounding regions thoroughly mapped and explored, even with the Arctic widely inhabited and the Antarctic host to permanent scientific stations—both regions remain, for most, places of mystery. Inasmuch as they are considered at all, they are viewed as remote, cold, barren, forbidding, and inhospitable. And so they are, in many ways. But they have played, and continue to play, a surprisingly prominent role in the natural and human history of the rest of the world.

The polar regions exert a profound influence on global climate: blasts of Arctic and Antarctic air bring cold to the Northern and Southern Hemispheres, respectively, and the melting of ice and sinking of cold, dense water in the polar oceans give a critical kick-start to the engine that drives the system of ocean currents worldwide. In the eighteenth and nineteenth centuries, the pelts of marine mammals from both regions sustained a massive industry that brought riches to places as far afield as England and China, and this in turn proved a catalyst for a series of voyages that lifted the veil from some of the mysteries of the polar realms. Oil from whales killed in the Arctic and Antarctic fueled the lamps that lit the streets of London, helped grease the wheels that turned the Industrial Revolution, and found uses in a suite of applications from soap to nitroglycerine. War has been fought over islands in the subantarctic. World War II was waged partly in the Arctic; World War III, it was long assumed, would be. The Arctic contains by far the largest oil fields in the United States and among the largest in the world; whether or not protected areas of northern Alaska should be opened to oil drilling has been a contentious political issue for at least twenty years, and it seems likely to remain so for some time to come.

But just as the polar regions have influenced the rest of the world, the world has left its mark on the poles and their environs. The fur pelts that brought riches to faraway lands came from the backs of fur seals and sea otters, whose populations were decimated by the slaughter. Similarly, the whales that surrendered oil and other products were, at both ends of the Earth, almost wiped out. Although oil exploration and development have been halted, perhaps forever, in Antarctica, they continue apace in the northern realms, sometimes causing pollution of both Arctic and subarctic environments. More recently, the image of an "ozone hole" above Antarctica and the specter of melting ice caps and rising sea levels as a result of global warming have brought home the potential consequences of human industrialization for even the most remote and pristine environments.

The story of the Arctic and the Antarctic is of two regions that are quite unlike any other—but that are almost as different from each other as each is from the rest of the world. It is a story of interweaving cycles in which exploration leads to exploitation, and exploitation to further exploration. It is a story of how even such remote realms can significantly affect, and in turn be deeply influenced by, events and trends thousands of miles distant—of how the long shadow of humanity has extended, for better and for worse, to the very ends of the Earth.

Chapter 1

Poles Apart

Southern Ocean, December 1994

There was a knock at the door and a flood of light from the passageway into my windowless cabin. One of the crew stood by the side of my bunk, a cup of tea in her hand. She placed the cup on a ledge and smiled as she looked at my bleary, half-comprehending face.

"There's a penguin outside, in a uniform and holding a Welcome sign," she said. "We must be in Antarctica."

She had laughed the night before when I asked to be woken when we crossed the Antarctic Convergence. There are, after all, no official markers delineating the Antarctic Regions' boundaries, and if the Convergence is shown at all on maps, it is typically as a vague, wavy line. But despite its ephemeral nature, the Antarctic Convergence is a very real, and very noticeable, border. I had never missed its crossing on any of my previous trips to the region, and I was not about to do so now, even if it meant clambering out of bed at 4:00 A.M.

The Convergence is where the cold waters of the Antarctic meet the temperate waters of the Atlantic, Pacific, and Indian Oceans. Heading toward the Convergence from the north, at first you see little sign of the shift that is about to take place. Then, over the course of a few hours of steaming or sailing, everything begins to change. The water temperature drops by several degrees. The air becomes noticeably more chilly. Where warm and cold air clash, a wall of mist rises to envelop you. Continuing on through the fog, you emerge on the other side in another world: The bird life is different. You may start seeing pieces of ice drifting on the water's surface. Far ahead, the pack ice, unseen but reflecting light, can create a phenomenon known as "ice blink," causing the horizon to appear brighter in the south than in the direction from which you have come. Maybe an iceberg will drift by. Having reached this far north, the iceberg is most likely old and weathered and melting rapidly. Waves may have carved arches and caverns into its expanse, perhaps revealing a bright cobalt blue interior. Standing on the ship's deck—or, by now, looking out from the protective warmth of the wheelhouse—you watch it slip silently past, and you realize you are now in the Antarctic.

There are those who claim that the real Antarctic does not begin until the vicinity of the Antarctic Circle, several hundred miles to the south. The Circle encloses almost all of the Antarctic Continent; it is the line south of which the sun does not set at the peak of summer or rise during the dead of winter. And certainly, those few areas of Antarctica that are north of this line—particularly the extension known as the Antarctic Peninsula, which reaches up toward the tip of South America—are frequently different in feel and form from those farther south. But the Convergence is the unquestioned portal, the opening not only to the continent itself but also to the ocean that surrounds it. Even as it moves slightly, its basic track remains the same from year to year, molded by ocean currents and the location of landmasses: after slicing through the Drake Passage, the channel between the Antarctic Peninsula and Tierra del Fuego, it arches north to around 50°S latitude as it heads into the South Atlantic. It more or less maintains that latitude around the Indian Ocean side of Antarctica before dipping lower, south of 60°S latitude, on the approach to Australia. It keeps south of 60°S as if to be certain of giving Australia and New Zealand an extra-wide berth—averaging a distance of perhaps 750 miles from them both—and then bulges up one more time and, finally, dips back through the Drake Passage.

South of the Antarctic Convergence lie 14 million square miles of ocean, one-tenth of all the ocean on the planet and some of the stormiest seas in the world. Antarctica's remoteness—it is 2,500 miles from Africa, 1,500 miles from Australia, and more than 450 miles from the closest land, the southern tip of South America—helped keep human explorers at bay until the nineteenth century. In the years before then, the tempestuous waters encircling the continent served to deter the few who ventured that far south. Uninterrupted by landfall, circling endlessly and driven by powerful winds from the west, the Antarctic Circumpolar Current, the only true global current, propels a seemingly endless procession of low-pressure systems through the Southern Ocean. These systems bring with them furious, screaming winds and waves of often terrifying dimensions: towering, whitecapped walls of water that can toss and toy with any vessel bold enough to venture into their domain.

Virtually the entire global landmass is north of the Antarctic Convergence. With the exception of a few scattered islands, only one piece of land lies within its boundaries—Antarctica itself, the highest and lowest, driest, and coldest continent in the world.

It is the highest because of the enormous ice sheet that blankets all but a few parts of the continent and boasts an average depth of 6,285 feet, or more than one and a quarter miles. Add the landmass that lies beneath and Antarctica's average height above sea level is around 7,800 feet far higher than the runner-up, Asia, which crosses the line at a mere 3,000 feet or so. It is the lowest because the sheer weight of all this ice presses down on the bedrock to such an extent that the peaks of many of the continent's mountains, though Himalayan in scale, barely clear sea level. It has been estimated that were the ice sheet to suddenly disappear, the greater part of the continental landmass would rise up higher than 3,000 feet: more than half a mile.

It is the driest because there is very little precipitation: there is perhaps four inches of snowfall each year over the polar plateau, maybe twenty inches in the coastal region (in comparison, Boston receives an average of forty-two inches annually), and virtually no rainfall. And, not surprisingly, it is the coldest. (The Arctic, though somewhat warmer, is disqualified from this competition because it is not, as we shall see, a continental landmass.) Outside of the Antarctic Peninsula, which some hard-liners sniff is not the "real" Antarctica, the highest temperature ever recorded was about 48°F. At the South Pole, the average annual temperature is about −60°F. The coldest temperature ever

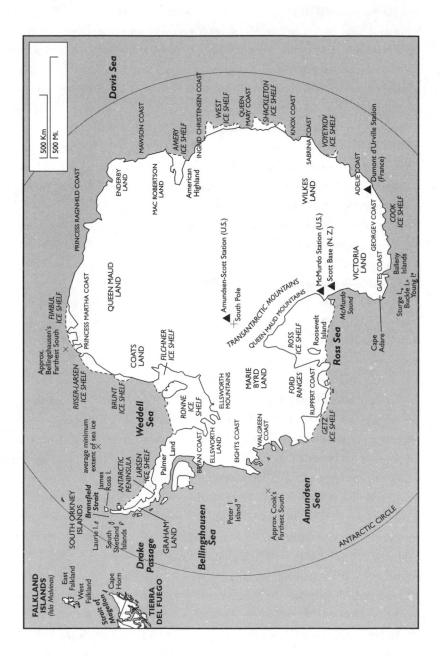

FALKLAND
ISLANDS
(Isla Malvinas)

East
Falkland
West
Falkland

Cape
Horn

Strait of
Magellan

TIERRA
DEL FUEGO

SOUTH ORKNEY
ISLANDS

Laurie I.

South
Shetland
Islands

Drake
Passage

Bellingshausen
Sea

GRAHAM
LAND

Bransfield
Strait

ANTARCTIC
PENINSULA

James
Ross I.

Palmer

LARSEN
ICE SHELF

average minimum
extent of sea ice

Peter I
Island

Approx. Cook's
Farthest South

BRYAN COAST

ELLSWORTH
LAND

EIGHTS COAST

WALGREEN
COAST

GETZ
ICE SHELF

Amundsen
Sea

ANTARCTIC CIRCLE

RIISER-LARSEN
ICE SHELF

Approx.
Bellingshausen's
Farthest South

BRUNT
ICE SHELF

RONNE
ICE SHELF

Weddell
Sea

COATS
LAND

FILCHNER
ICE SHELF

ELLSWORTH
MOUNTAINS

MARIE
BYRD
LAND

FORD
RANGES

RUPERT COAST

FIMBUL
ICE SHELF

PRINCESS MARTHA COAST

QUEEN MAUD
LAND

TRANSANTARCTIC MOUNTAINS

QUEEN MAUD MOUNTAINS

ROSS
ICE SHELF

Roosevelt
Island

Amundsen-Scott Station (U.S.)

South Pole

Ross Sea

McMurdo
Sound

Cape
Adare

Davis Sea

PRINCESS RAGNHILD COAST

ENDERBY
LAND

MAC ROBERTSON
LAND

MAWSON COAST

AMERY
ICE SHELF

American
Highland

INGRID CHRISTENSEN COAST

WEST
ICE SHELF

QUEEN
MARY COAST

SHACKLETON
ICE SHELF

KNOX COAST

SABRINA COAST

WILKES
LAND

VICTORIA
LAND

McMurdo Station (U.S.)

Scott Base (N.Z.)

GATES COAST

GEORGE V COAST

VOYEYKOV
ICE SHELF

ADELIE COAST

Dumont d'Urville Station
(France)

COOK
ICE SHELF

Balleny
Islands

Sturge I.
Buckle I.
Young I.

500 Km.

500 Mi.

recorded in Antarctica—the coldest ever recorded on the surface of Earth—
was −129.9°F.

Because of their positions at the ends of the world, the polar regions vary
between periods of prolonged, almost continuous, sunlight during summer and
months of protracted darkness during winter. This is because Earth tilts as it
moves around the sun, offering alternately the Northern and Southern Hemispheres to the star it orbits. This is what gives Earth its seasons; it is why summer in the Northern Hemisphere is winter south of the equator and why the
equatorial zones receive almost unchanging amounts of sunlight over the
course of the year. But because the Arctic and Antarctic are so far north and
south, respectively, the angle of the sun to Earth is such that even in summer,
the sun never rises as high above the horizon as it does over the equatorial zone.
As a result, observes ecologist Bernard Stonehouse, polar sunshine "can be
strong enough to warm rocks, melt snow and encourage judicious sunbathing;
it contains enough ultraviolet radiation to cause sun-tanning, serious sunburn
and snow-blindness, but it cannot bring lasting warmth to polar areas."

Antarctica, then, is so cold because of its location and its isolation; at the
same time, its blanket of ice, itself a consequence of the extreme conditions,
contributes to the continuation of these extremes. What little solar energy
reaches the region is reflected back into space by the ice cap. And although
there is some circulation of warm waters from north of the Convergence, the
Antarctic is so distant from the rest of the world, and the barrier formed by the
Convergence and polar currents so effective, that the warming effects of tropical waters are essentially unnoticeable. Whereas the Arctic experiences a
degree of interchange between polar and temperate weather systems, the
Antarctic is all but sealed in its own self-perpetuating cocoon of cold.

Antarctica is the fifth largest continent, one and a half times larger than
Australia, with an area greater than that of western Europe, larger than the
United States and Mexico combined. Of its 5.4 million square miles, roughly
10 percent is taken up by ice shelves, large, semipermanent areas of ice that are
anchored to the mainland. Fed by glaciers and ice streams—rivers of ice that,
apparently because they rest on foundations of water-saturated sediment, move
forward at a greater pace than the ice sheets surrounding them—these shelves
are forever varying in size and shape and constantly calving off icebergs, on

occasion tens of miles in length. Nonetheless, despite their comparative incon-
stancy, they are an integral part of the Antarctic continent.

Surrounding the continent is a permanent ring of sea ice. In summer, this
sea ice covers about 1.5 million square miles, but at the height of winter it
expands to blanket some 7.5 million square miles. Accordingly, the total
expanse of Antarctica, sea ice included, all but doubles from summer to winter, from around 7 million to 13 million square miles.

Whereas ice shelves and the icebergs they calve are freshwater, vast accu-
mulations of millennia of snow, sea ice is the frozen surface of the sea and,
accordingly, salt water. Much of the salt content of sea ice gradually becomes
concentrated at the bottom of the floes, and from there it is extruded back into
the sea. Thus, all but the freshest sea ice is less salty than the seawater from
which it was formed. As a result, in spring, its melting—and the melting of the
icebergs that split from the ice shelves—dilutes the surface water of the Southern Ocean. This near-freezing water slowly spreads north, its departure com-
pensated for by warm water from the Atlantic, Pacific, and Indian Oceans flowing underneath the Convergence to the edge of the Antarctic continental shelf.
There, bringing with it a cornucopia of nutrients, this warmer water rises to
the surface, forming what is known as the Antarctic Divergence.

Some of this water, now cooler than before, heads north again toward the
Antarctic Convergence. The rest of it continues south until it reaches the edge
of the continent itself. Its temperature lowered by contact with the ice shelves,
it eventually sinks to the bottom and suffuses across the seafloor, slowly, inex-
orably spreading outward and northward again. It continues on, beneath and
past the Convergence, into the temperate waters of the rest of the world's
oceans, all the while hugging the seafloor. It is a journey that takes decades or
even centuries, and because of it, much of the global ocean is a couple of
degrees cooler than it otherwise would be. Although concern over potential
adverse effects of global warming on the Antarctic has tended to highlight the
prospect of the Antarctic ice sheet collapsing and causing a massive increase in
sea levels, such a development is widely considered by scientists to be highly
improbable, at least for several hundred years. More likely, although still
requiring an increase in temperatures beyond that thus far experienced or predicted, is a melting of the ice shelves and possible interruption of this "conveyor

belt" of cold ocean currents, with uncertain consequences for the rest of the world.

The sea ice is a boon for Antarctic life. It provides a platform for algae, tiny microscopic plants, some of which become embedded in the ice as it forms. In spring, as the ice begins to melt, pigments in the algae may absorb solar radiation and so cause the ice to melt faster. As it melts, the algae are released into the water, where they are fed upon by tiny animals known as zooplankton. These in turn act as magnets for species progressively higher on the food chain.

Behind the intimidating barrier of the Antarctic Convergence, the Southern Ocean is home to relatively few species. Many of those species, however, may be found in surprisingly high numbers. There are, for example, just thirty-nine nesting bird species south of the Convergence—fewer, notes writer and researcher David Campbell in his book *The Crystal Desert,* "than can be found in a small garden in Colombia or Costa Rica." But the number of individual birds found around the coasts, islands, and seas of Antarctica is perhaps 70 million. As with birds, so with other wildlife. There are but four species of so-called true seals in the Antarctic. But according to some estimates, more than one in every two seals in the world is a crabeater seal, a species confined to the Antarctic; the total biomass (essentially, the combined weight) of crabeaters is more than four times that of all other seals and sea lions on the planet combined.

The distribution of this natural abundance is patchy, confined to oases in what is otherwise a relatively empty and forbidding expanse. Campbell observes, for example, that just one site, "volcanic Zavadovskiy Island, in the South Sandwich archipelago, has a colony of approximately 3.5 million chinstrap penguins." As anyone who has been in their midst can testify, penguin rookeries are densely packed, noisy, and above all smelly affairs, with little black-and-white birds stretching seemingly to the horizon. Outside such colonies, though, penguins are rare: if they were spread out evenly throughout the Southern Ocean, there would be just 11 ounces of penguin per square mile.

As in any marine ecosystem, microscopic algae—phytoplankton—in the Southern Ocean sit at the base of the food pyramid. The dominant, distinguishing feature of the Antarctic seas, however, is the zooplankton—especially the small shrimplike creatures called krill. As much as two inches long, krill are, by zooplankton standards, massive. Their significance, though, lies less in their individual size than in their habit of congregating in swarms, huge concentrations that can cover thousands of square yards of ocean and color the sea red. A krill swarm can extend to depths of roughly 130 to 160 feet, contain thousands of krill per square yard, and weigh as much as 2.5 million tons, although most are smaller.

Seals, penguins, and many seabirds other than penguins depend on krill as their primary source of food. So, too, do the Antarctic's largest denizens, the mighty baleen whales. The blue whale, the largest of all whales—indeed, the largest animal ever to live on Earth—is found in the Antarctic, where it spends the southern summers feeding before heading north to breed in warmer waters during polar winter. Despite their enormous size, blue whales in the Southern Ocean feed almost exclusively on krill, as do their relatively smaller cousins, including the fin, sei, humpback, minke, and right whales. This they do by swallowing huge amounts of water and forcing it through their sievelike baleens—long, fringed plates made of keratin (the same substance as in our fingernails) that hang from their upper jaws. The baleen filters out the water and leaves the krill, which the whales swallow in gargantuan mouthfuls.

At least in theory, most Antarctic wildlife is now strictly protected, but it has not always been so. As Ernest Shackleton and his men drifted north on their ice floes after being forced to abandon their ship, they survived to a large extent by killing seals and eating their meat, and the stack of dead penguins that remains even today outside Robert Falcon Scott's hut in Antarctica speaks to the importance of the birds' flesh to that expedition. The effect of such hunting on wildlife numbers in the region was insignificant; that of the slaughter visited on whales and fur seals—not by explorers but by commercial hunters from the north—was not.

More recently, some have salivated at the prospect of large-scale commercial exploitation of krill, but outside of a few markets, the projected demand for the crustacean as a source of protein has not really emerged. Krill have a strong flavor, and a taste for them is not easily acquired; their meat contains high levels of fluoride; and the krill themselves easily spoil. Instead, krill are valued more for their carotenoids, chemicals that give krill their coloring and that, mixed with feed for farm-raised salmon, help ensure that the fish attain the pinkish hue consumers demand. Although krill fishing has not reached the scale once anticipated, some Antarctic fish species have been subjected to sur-prisingly high levels of exploitation—not by those who visit the region as explorers and researchers but by commercial fishers working on behalf of those same consumers in faraway lands. The fish widely marketed in the United States as Chilean sea bass, for example, is known to scientists as the Patagonian toothfish, and its popularity on dinner menus has brought the species—as was the case with Antarctic cod before it—to the brink of extinction.

There are perhaps 200 fish species south of the Antarctic Convergence, a mere 1 percent of all fish species in the world, and most of them are endemic, unique to Antarctic waters. Unlike seals and penguins, Antarctic fish generally do not make up for their lack of variety through sheer numbers of individual species. Indeed, many have had to develop unusual qualities simply to survive in the region's frigid waters. Seawater freezes at lower temperatures than freshwater because of the salts it contains, but fish blood and tissues contain lower concentrations of such salts and so tend to freeze at higher temperatures. In most parts of the world, that is not a problem; in the frigid Antarctic, how-ever, it is. In response, many fish species in the Southern Ocean have evolved a kind of antifreeze, a mixture of sugars and proteins that interferes with the for-mation of ice crystals in their blood and lowers the temperature at which the blood freezes by about 3.6°F, enabling the fish to function even as the water above them turns to ice.

Through such adaptations, the flora and fauna of the Antarctic stand as a remarkable tribute to the versatility and resilience of life. As in the ocean, so on the continent itself: communities of tiny lichens—fungi and algae that have evolved together in a symbiotic relationship—actually live inside rock. Hard-shelled mites have been found living on nunataks—mountain peaks that poke above the ice sheet—as close to the South Pole as 85°S latitude. And researchers in 1999 reported finding bacteria, apparently respiring, about 11,000 feet (more than two miles) deep in the ice. Even so, in all but the most northerly areas, life struggles to maintain its foothold. The largest year-round inhabitant of the continent is a wingless midge less than half an inch long; much of Antarctica's terrestrial life is even smaller. There are only two species of flowering plant, both of them in the northerly Antarctic Peninsula region. The mosses and lichens that grow in the Antarctic do so slowly, adding only about

0.6 inch every 100 years or so, which makes them highly vulnerable to the effects of human footsteps.

Such footsteps were a particularly long time in coming to the Antarctic, and today they remain infrequent. The first people to cross south of the Antarctic Circle did so only in 1773. The continent was not sighted until 1820, and the first confirmed landing was not until 1895. The South Pole was attained for the first time in December 1911 and for the second time the following month; no one stood there again until 1956. There are no permanent human inhabitants on the frozen continent, merely visitors—scientists and support staff, around 3,500 in summer and less than half that in winter. Scattered among forty or so different bases, they seek to uncover the mysteries the Antarctic clasps to its icy bosom. Technology has improved the comfort level of clothing and shelter for those who choose to spend weeks or months at a time in Antarctica, but it cannot remove the dangers of a land where a violent snowstorm can spring up at any moment, causing anyone caught outside to become lost and in danger of dying even within yards of camp.

And so it will quite likely always remain. Humans are, at best, intruders here, cautious interlopers into one of the most hostile environments on the planet. At the other end of the Earth, it is a somewhat different story. There, too, conditions can be hostile and even deadly, so much so that the Arctic is still among the world's least visited and most sparsely inhabited places. Inhabited by humans it is, however, and it has been so for thousands of years.

Off the northern coast of Alaska, August 1998
It was a Sunday morning, with little to be done except the work needed around our ship. Arne, the captain, was on the bridge when, some way in the distance, he saw something swimming toward us.

Closer inspection revealed a polar bear swimming steadily in our direction, hind legs tucked under it, front legs paddling away, head held high. Closer and closer it came until it was just a matter of feet off the stern, swimming back and forth, checking us out. In the middle of the Beaufort Sea, it could smell cooking, and it was looking for lunch. But lunch, refusing to cooperate, was walking around on deck looking back at it. Tempting as it may have been, populated with unusual vertical seals, this big green ice floe clearly was not giving up its contents anytime soon. And so, reluctantly admitting defeat, the bear turned away from us, and we watched as it slowly swam into the distance.

A wild polar bear, whether swimming or swaggering along the edge of an Arctic ice pack, is an awe-inspiring sight. Even though our crew had seen many natural wonders around the world, a cry from the wheelhouse that a polar bear had been sighted was almost invariably enough to send everybody rushing up on deck. That we were able to see any, let alone so many, was a privilege not to be taken for granted: until the countries that surround the Arctic Ocean agreed to impose international protective measures in 1973, the polar bear's survival appeared in jeopardy. Unlike Antarctica, the Arctic has been host to human populations for many hundreds of years. Particularly as the number of visitors and settlers increased over the twentieth century, the polar bear was among the species that most suffered from the intrusion. Its thick pelt and ample meat supplies were tempting to embattled explorers and commercial traders alike; frequently, however, the shooting of polar bears was motivated largely by fear. Polar bear attacks on humans are surprisingly rare, but fear of the prospect is not hard to understand. Standing nearly ten feet tall and weighing more than 1,400 pounds, polar bears are the largest of all bears and an intimidating presence.

Arguably the iconic species of the Arctic, polar bears do not exist in the Antarctic; cartoon Christmas cards notwithstanding, polar bears and penguins never coexist. Nor could they ever: although there was once an aborted and ill-advised effort to introduce penguins to the Lofoten Islands, off the coast of Norway, it did not succeed. In the Antarctic, polar bears might at first find enough food to sustain them for a short period in the summer months, but assuming they were able to weather the extreme Antarctic winters, they would eliminate much of the continent's penguin population in a matter of a few years.

Penguins live in the Antarctic;* polar bears live in the Arctic. It is an easy distinction to remember, but it is just one of many. The Arctic and Antarctic do

---

*Among other places—including such non-Antarctic locales as the Galapagos Islands, where they are nurtured by cold-water ocean currents. But not in the Arctic—or for that matter anywhere close.

share certain traits. Courtesy of their relative positions at either end of the
world, both enjoy almost uninterrupted sunlight during the summer and
endure seemingly endless nights in winter. Both are, particularly in winter,
very, very cold. But there are at least as many differences as similarities.

The Antarctic is isolated from the rest of the world. The Arctic, far from
isolated, includes parts of three continents and the largest island on the planet.
Political boundaries, nonexistent or meaningless in the Antarctic, have real significance in the Arctic. In the Bering Strait, for example, Little Diomede Island
is a village in Alaska, the forty-ninth state of the United States. Just a couple of
miles away, easily visible and in theory just as easily reachable, is Big Diomede
Island. But Big Diomede is Russian territory and located on the other side of
the International Date Line. Almost close enough to touch, it is in a sense
another day—and, at least during the cold war, a whole world—away.

Antarctica is a continent of ice surrounded by ocean. Assuming you could
find a way to get there, the South Pole is easy to identify. There is a U.S.
research station, and an actual pole, to mark the spot. Stand there and you are
supported by solid ice almost two miles thick. The Arctic, by contrast, is a
frozen ocean encircled by land. There is no marker to identify the North Pole.
There cannot be: it is in the middle of pack ice that is constantly shifting, being
pushed by currents in the direction of Scandinavia. Although the pack is thick
enough to stand on and walk over—and several large floes support temporary,
floating research stations—it is also sufficiently permeable that icebreakers
replete with tourists can force their way to the Pole and nuclear-powered submarines can break through from below.

Both the Arctic and the Antarctic boast year-round ice, but there is far less
of it in Arctic lands than on the southern continent. Ninety-eight percent of
Earth's land ice is in the polar regions, but 91 percent is in Antarctica. Most of
the rest is taken up by just one part of the Arctic—Greenland, which is covered by the largest ice sheet outside the Antarctic. Thus, the Arctic is not super-
cooled by an ice cap, as Antarctica is. And without a barrier like the Antarctic
Convergence, the Arctic is more susceptible to the influence of warm currents
and weather systems originating outside the region. As a result, Arctic and subarctic locations are generally much warmer than those at a similar latitude in
the Southern Hemisphere. Helsinki, the subarctic capital of Finland, at a lati-

tude of 60°N, basks in mean midsummer temperatures of 63°F, some 32 degrees warmer than those in summer on Signy Island, an Antarctic research station at the equivalent southern latitude.

As a much more hospitable environment than the Antarctic, the Arctic is blessed with a far greater abundance of species. For example, whereas the entire continent of Antarctica is host to only 2 flowering plant species, Greenland alone boasts more than 40. Devon Island, a northern outpost in the Canadian Arctic, is on the boundary between arctic tundra and polar desert—the environment of the so-called High Arctic, perhaps the nearest Arctic equivalent to the desolate plateau of Antarctica. Even on its lowlands, which represent the poorest kinds of tundra, botanists have found more than 130 species of moss, 30 species of liverwort, 9 species of fern, and 90 species of flowering plant, all within the space of a few acres. Arctic tundra supports 8 nonmigratory land bird species, a trifle compared with, say, the numbers in the Amazon rain forest, but 8 more than in Antarctica. All together, the Antarctic (including its outlying islands) has 39 nesting bird species; the Arctic has 150. Antarctica has no land mammal species; the Arctic has 40, including relative giants such as moose, musk oxen, and bears. And, of course, the Arctic has been home to self-sufficient human communities for more than two thousand years, whereas no expedition to Antarctica has been able to survive entirely on regional resources, and it is unlikely that one ever could.

Nonetheless, the Arctic retains a reputation of barren desolation. English missionary S. K. Hutton, writing early in the twentieth century, described the interior of northern Labrador as a "bare and desolate waste, silent but . . . for the dismal howling of the hungry wolf, or the even more dismal howling of the wind." That assessment has held sway in much of the public consciousness, despite frequent contradictions by subsequent observers. Perhaps the most vocal of those later cheerleaders of the region's riches was writer and explorer Vilhjalmur Stefansson, who sarcastically noted that the Arctic "is lifeless, except for millions of caribou and foxes, tens of thousands of wolves and muskoxen, thousands of polar bears, millions of birds, and billions of insects." Stefansson characterized the region as the "friendly Arctic," but his view also has been criticized, for overstating the region's bounteousness. After all, in terms of weight, a tundra area produces only about 1 percent of the plant life of an area of sim-

ilar size in the temperate Northern Hemisphere. And although native peoples
have thrived on the Arctic's riches for thousands of years and many people live
there still, the temperatures regularly drop low enough that without adequate
clothing and shelter, a person would die within fifteen minutes.

A more accurate and nuanced vision is conveyed by archaeologist William
Taylor:

> Southerners commonly think the immense circumpolar world
> is remote, empty, cold, hostile and lethal. It sometimes is, but
> so are huge southern cities. Although it can be a forbidding
> moonscape, the Arctic is also varied, majestic, serene, memo-
> rably beautiful and occasionally gentle. The far north is not only
> a prowling bear, a battering storm and vicious cold, but also a
> fat bumblebee buzzing among delicately yellow arctic poppies.

Opinions vary as to how exactly the Arctic should be defined. Latitude
alone will not do it. The Arctic Circle, for example, excludes Iceland and large
swaths of Siberia and includes, in the words of writer Barry Lopez, "a part of
Scandinavia so warmed by the remnants of the Gulf Stream that it harbors
a lizard, an adder, and a frog." Labrador, which filled the aforementioned
S. K. Hutton with such dread, is at best on the Arctic's fringes. Berlin, in Germany, is plainly not in the Arctic or even the subarctic, but it is on the same latitude as Irkutsk, Siberia, which—with winter temperatures around –40°F—
arguably is.

Not surprisingly, therefore, it is difficult to put a figure on the Arctic's size,
although most estimates place it somewhere between 11 million and 14.5 million square miles of land and sea. At the center of it all, the Arctic Ocean itself
is generally considered to be the area of ocean that is more or less permanently
covered by sea ice, an area of about 5.66 million square miles. Some oceanog-
raphers also grant true Arctic status to waters such as those of Hudson Bay,
which, although not permanently frozen and despite their temperate latitude,
mix almost entirely with waters from the Arctic Ocean. But in terms of climatic conditions and the wildlife they support, marine areas outside that limited range—such as Baffin Bay, Davis Strait, and the Beaufort, Bering, Laptev,
and Kara Seas—also lay claim to Arctic qualities. Even if they are not perma-

nently frozen, all these areas remain at least partly frozen for some of the year; therefore, many scientists accept the southern limit of the winter pack ice as a good line for determining the maritime Arctic. By this measure, Arctic seas include, as well as those just mentioned, Hudson Bay, Hudson Strait, and parts of the Labrador Sea, the Barents Sea, and the Sea of Okhotsk. They also include islands such as Greenland—which is Arctic by almost any definition—Novaya Zemlya, Svalbard, and Baffin, Banks, and Victoria Islands.

The Arctic Ocean basin freezes relatively easily, not only because of its location but also because it is relatively shallow (on average about 5,010 feet, a little less than a mile, deep) and because large river systems from Alaska and Siberia drain into it and there is little water exchange with other seas, causing the salt content to be lower than in other oceans. The result is the largest expanse of floating ice in the world. This sea ice cover ensures that in Arctic regions, the division between terrestrial and marine environments is, at least in certain places and at certain times, uniquely blurred. Polar bears den and give birth on land, prefer to live and hunt on the ice, and are so at home in the sea that they are officially classified as marine mammals. On very rare occasions, the ice may even come ashore: large, fast-moving floes driven by currents and winds may pile up along the beach until suddenly they break over the land barrier and grind their way rapidly across beach, sand spit, or island. Eskimos call this phenomenon *ivu*. At Point Barrow, the most northerly part of Alaska, archaeologists found the remains of a house and its occupants who were suddenly overwhelmed by *ivu*, their frozen bodies testimony to its force and rapidity.

Roughly shadowing the southern limit of pack ice, but in places extending a little farther south to take in the Norwegian Sea, virtually all of the Barents and Bering Seas, and most of Iceland, is an invisible line known as the 10°C (50°F) July isotherm. This line delineates a zone where the temperature in the warmest month of the year, normally July, reaches 10°C, or 50°F. By some definitions, the area north of this line is the Arctic and that immediately to the south is the subarctic. Although this may seem a little arbitrary, it does approximately correlate with what is perhaps the most widely accepted ecological boundary between the terrestrial Arctic and subarctic: the tree line, or timberline, where boreal forest gives way to tundra.

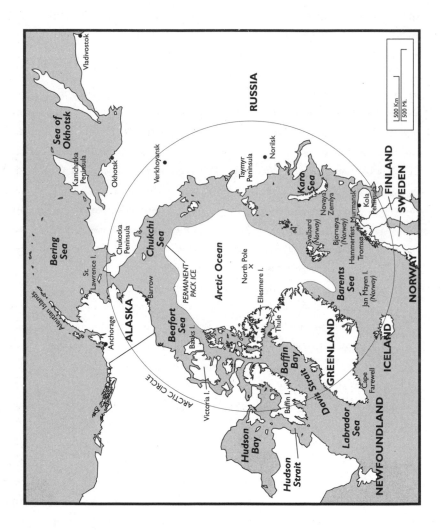

Boreal forest is the principal ecosystem of the subarctic, forming a band that covers most of Alaska, more than half of Canada, much of Scandinavia, and approximately 3 million square miles of Russia. Dominated by conifers—depending on location, primarily spruces, larches, and pines—it is interspersed with deciduous trees, shrubs, bogs, ice fields and glaciers, rivers and streams, lakes and alpine tundra. At its deepest points inland and thus farthest removed from what even in the Arctic is the warming effect of the ocean, it can be far colder in winter than the North Pole and its environs many hundreds of miles to the north. The record for lowest temperature in the Northern Hemisphere, −90°F, is held by Verkhoyansk, a Siberian town just north of the Arctic Circle.

The forest is sometimes known as taiga (pronounced *tie*-ga), a misappropriation of a Russian word describing a landscape in which sparse coniferous forest is interspersed with scrub, peatlands, and treeless barrens. More accurately, the term *taiga* should be confined to the northerly edges of the forest, the point where conditions begin to be too harsh for trees to survive. This is the tree line—not a sharp boundary at all but a place where conifers become steadily more sparse and tundra progressively more commonplace. North of the tree line lies the vast expanse of tundra and, by many definitions, the true terrestrial Arctic.

The term *tundra* comes from a Finnish word referring to open barrens and dwarf woodlands, something not far from the true definition of taiga, with which it is contrasted. Confusingly, it is used to describe both a specific type of vegetation and a geographic area in which that vegetation predominates. In the geographic sense, it generally means the open, treeless region north of the boreal forest and south of the "polar desert" of the High Arctic. The form of vegetation to which it applies, which is by no means confined to that same geographic region, is typically a low-standing mixture of shrubs, grasses, flowering plants, mosses, lichens, and algae.

Like the Arctic itself, the tundra has earned something of an unflattering reputation. Perceived as barren and bleak, it has become synonymous with a region of virtually no vegetation and little life. Far to the north, in the High Arctic, that assessment becomes progressively more accurate: north of the 6°C (42°F) summer isotherm, the tundra becomes thinner and vegetation becomes more stunted, rarely more than eight inches high; beyond the 2°C (36°F) isotherm, only the toughest lichens, mosses, algae, and tiny flowering plants remain. Conditions on the tundra of the Low Arctic can also be uninviting. It is covered with snow in winter, frozen solid for eight months of the year. When the snow melts in spring, a layer of permanently frozen soil, the permafrost, prevents meltwater from draining away, creating a boggy patchwork.

But during the brief summer the region comes alive, bursting forth in a mosaic of colorful flowers. Bumblebees buzz from bloom to bloom. Clouds of alarmingly persistent mosquitoes fill the air. Summer visitors join the few hardy birds and mammals—ptarmigans, ravens, snowy owls, lemmings, voles, and others—that have braved the winter months. Multitudes of birds, from large cacophonies of geese, swans, ducks, and other shorebirds to gulls, terns, and smaller waterfowl, crowd onto the tundra and the coastal river deltas from their wintering points far off. Caribou (in North America) and reindeer (in Eurasia) begin massive migrations millions of animals strong, sweeping up from the tree line to the coastal plains.

In the seas, too, the onset of spring and summer marks the start of a series of migrations. Walruses, belugas (white whales), and seals thread their way through the leads that open up in the winter pack ice. In the western Arctic, bowhead and gray whales swim north via the Bering Strait toward Point Barrow and beyond. Like the caribou migration on land, this northern passage of marine wildlife can be a spectacular sight: Cape Sarichef, on Unimak Island in Alaska's Aleutian Islands chain, provides an unmatchable vantage point for watching the sea mammals, which must travel through the narrow passage of Unimak Pass to continue to the Bering Sea.

Suddenly, the Arctic is no longer a forbidding and near-lifeless realm. Seemingly at once, it is transformed from a frigid frontier to a region of plenty. And it is such plenty that enables the continued presence of another of the Arctic's species, one of its more recent arrivals, scattered throughout the region from the cold interior of the Russian North to the coasts of Greenland: representatives of the most numerous and widespread large mammal in the world.

Chapter 2

Hunting the Bowhead

They call themselves simply "the people," or, sometimes, "the real people." On Alaska's western coast, they are the Yup'ik. On Russia's Chukotka Peninsula and on St. Lawrence Island in the Bering Sea, they are Siberian Yup'ik. In northern and northwestern Alaska, they are Iñupiat; in the western Canadian Arctic, Inuvialuit; in central and eastern Arctic Canada and in Greenland, they are collectively known as Inuit.

The first westerners to encounter them cared little for such distinctions and called them all Eskimos, a name derived, according to some scholars, from a term used by the Montagnais Indians, an aboriginal people of what is now Quebec. This Montagnais word, *ayassimew,* may mean something like "snow-shoe-netter," but in due course, it apparently came to be adopted by Algonquin Indians and transformed into a word that meant "eater of raw flesh." That is the interpretation that stuck, and because of its pejorative undertones, *Eskimo* has come to be widely shunned and *Inuit* substituted in its place. Nonetheless, the Yup'ik and Iñupiat continue to describe themselves in general terms as Eskimos, and anthropologists use the word to describe the particular type of culture that the Inuit, Inuvialuit, Yup'ik, and Iñupiat all share.

That culture has its roots in the Bering Strait, fifty-five miles of water and, in winter, sea ice, which is the closest point between Old World and New. The strait acts as a funnel for marine life, a bottleneck through which swim, during their migrations north and south or as part of their regular travels in search of food, bowhead and gray whales, belugas and narwhals (small whales, relatives of dolphins, with just two teeth, one of which, in males, breaks through the upper jaw and into a spiraling tusk as much as ten feet long), ringed seals, ribbon seals, spotted seals, bearded seals, and walruses. In spring and fall, the skies over the strait are filled with flocks of birds flying to and from North America and Siberia. The area is a natural magnet for a hunter-gatherer people, and there is evidence of subsistence cultures having been established along the Siberian coast at least 18,000 years ago, and quite possibly earlier.

There is no record of when the first adventurous souls set out eastward to investigate the land that, on a clear day, they could see on the horizon. It has generally been argued, however, that these pioneers did not row or sail across the water that divides Alaska and Siberia today. Instead, they walked.

Between 25,000 and 10,000 years ago, the Northern Hemisphere was in the grip of the last great ice age, and a greater percentage of the world's water was locked up in ice than is the case now. Accordingly, the sea levels of the Bering Strait were lower, about 330 feet lower than today, and the tundra of the Chukotka Peninsula extended east across what is now the Chukchi Sea. The result was the Bering Land Bridge, or Beringia, a wide swath of land connecting Asia and Alaska: an ice-free corridor that functioned as a refuge for many species of plants and wildlife, allowing a number of them to spread from Eurasia into North America. The circumpolar distribution of numerous species, from plants to large mammals such as the Eurasian reindeer and their North American cousins the caribou, is testament to the existence of the natural ark that was Beringia. And during that same period, perhaps pursuing such prey species as reindeer, musk oxen, or mammoths, perhaps seeking to escape the encroaching ice, perhaps both, *Homo sapiens* made the same journey.

The people who crossed Beringia may have been the first humans to arrive in North America. They were quite likely the ancestors of the native people who live in the interior of Alaska and, perhaps, of the Native Americans in the lower forty-eight. However, they did not, it seems, settle in the Arctic, and the presence of extensive ice sheets across Canada prevented them from heading east. They were not the forerunners of "the people."

But others kept coming, long after sea levels had risen and the land bridge had been swamped, rowing now across the returned stretch of water. And just as they had crossed the Bering Strait, they also crossed, when the climate allowed, the icebound islands of Canada's Arctic Archipelago and ultimately thence across the Nares Strait to Greenland. Over time, they developed new techniques for hunting, built different kinds of houses, and adapted to the changing climate and the wildlife they chose to hunt, some adopting a nomadic or seminomadic way of life, others basing themselves in somewhat permanent settlements.

This was not one great mass migration. Initially, hunters probably maintained a home base on the Siberian coast, venturing on forays to the other side of the strait in search of game. When they did settle, they did so in small groups, sometimes impermanent, sometimes numbering as few as hundreds or even tens of individuals.

Several waves of different cultures arose in the Bering Strait area, headed east, and then developed and changed or were supplanted by new arrivals. Here, on the coasts of Siberia and Alaska, there emerged around two thousand years ago a people known as the Old Bering Sea culture. They hunted both sea and land mammals from permanent coastal settlements; they traveled over the water by *qajaq* (kayak) and *umiaq* (walrus-hide boat), and they used dogsleds to traverse the ice. Other, similar cultures followed, including one on which archaeologists have bestowed the name of the settlement—Birnirk, near Barrow, on Alaska's northern coast—where researchers first uncovered evidence of its existence. The Birnirk people initially concentrated on hunting caribou and fishing, but apparently in response to changes in caribou numbers and distribution, they switched their attention almost exclusively to marine mammals, particularly whales. Theirs soon became the dominant culture of northern and northwestern Alaska; from them, we can draw a direct line to the present-day Iñupiat. The Birnirk people also spawned, at least in part, the Thule culture—the direct ancestors of most of the modern Inuit of Canada and Greenland—who from around 1000 to 1200 C.E. spread rapidly east, supplanting or absorbing existing cultures on their way as they took advantage of a warming climate and open pathways through the ice in pursuit of their quarry: the bowhead whale.

For their descendants—as for all peoples around the world—much has changed over the ensuing millennium. Modern-day Eskimos and Inuit live in prefabricated housing rather than sod dwellings; indeed, a plurality of Alaska Natives live not in villages at all but in Anchorage, a modern city of 200,000 people. They watch the National Football League and the World Wrestling Federation on satellite television and wear hooded sweatshirts emblazoned with NBA logos. They are as likely to settle down to a dinner of frozen pizza or macaroni and cheese as they are to hunt seals.

Nonetheless, in many instances, such changes are largely cosmetic. For the vast majority of the Inuit, Inuvialuit, Iñupiat, and Yup'ik, the central elements of their culture—their spiritual and nutritional well-being, their way of life, indeed their very existence—are the sea ice, the Arctic ocean that lies beyond, and the wildlife that both sea and ice support.

Throughout the Arctic, villagers hunt seals of all kinds—ringed, ribbon, spotted, and bearded. Depending on the location and the species involved, they may catch them by traveling in boats and looking for them along the ice edge or by adopting a technique similar to that used by polar bears: lying patiently near a breathing hole in the ice, waiting for a seal to poke its head out of the water to take a gulp of air, and then striking.

Yup'ik on St. Lawrence Island and in some communities in mainland Alaska, and Inuit in parts of northern Greenland, specialize in hunting walruses. In Canada and Greenland, Inuit hunt narwhals and their cousins, belugas. Each summer, off the Iñupiat village of Point Lay, the migration of several thousand belugas prompts the hunters of the village to set out in boats and herd perhaps forty or fifty of the white whales into the bay, where a year's supply of food is then secured in one day. In Greenland, some communities take minke whales. The Chukchi and Siberian Yup'ik peoples of the Chukotka Peninsula hunt gray whales as the whales migrate to and from the Sea of Cortés in Mexico. The Iñupiat, and the Siberian Yup'ik of St. Lawrence Island, hunt bowheads.

* * *

The bowhead is a massive chunk of whale. It is not nearly as long as the blue whale, or as heavy; nonetheless, with a length between 39 and 59 feet and a weight of as much as 120 tons, it is by any measure large. One-third of its length is taken up by its bow-shaped skull, which measures roughly 16 feet long and gives it its name. Its blubber, approximately 17 to 20 inches thick, exceeds that of any other animal. So massive is the bowhead, in fact, that it has perhaps the lowest ratio of surface area to body volume of any mammal in the world.

During the long winter, bowhead whales gather in polynyas—large ice-free areas—in the western Arctic and along the edges of the pack ice in the western and central Bering Sea. Toward the end of winter, as the ice starts to melt away, they move steadily north and east from the northern Bering Sea through the eastern Chukchi Sea along the coast of Alaska. Following leads in the ice, they swim past the western end of St. Lawrence Island from late March to early April; continuing north, they pass west of Big Diomede and head on toward Point Barrow. The first pulse of whales usually reaches Barrow in late April and early May, with a second pulse arriving in mid-May and a third and final one in late May to mid-June. For the Iñupiat, the whales' arrival marks the spring whaling season; but the bowheads do not linger. Instead, they swim some way offshore until they reach the eastern Beaufort Sea, and then they continue all the way to the Mackenzie River delta and on into Arctic Canada, where they spend their summers. By early September, they are on the move again, swimming back west through the Beaufort, closer to shore this time than during the spring migration, passing along Alaska's coast until early October. This time, too, the migration proceeds in pulses, beginning with juvenile whales, followed by cow–calf pairs, and ending with large, mature whales. Finally, they round the corner once more, heading back south into the Bering Sea to spend their winter.

The bowhead migration is the axis around which Iñupiat society revolves. The hunt itself lasts only about six weeks in spring and a month during fall, but activities surrounding it continue throughout the year. Beginning in October and November for the spring hunt and in June and July for the fall, villagers hunt bearded seals and walruses for skins to cover the *umiat*.* After being specially treated and stored, the skins are handed over to the village's women, who sew them together with caribou sinew.

Come spring, the crews begin constructing new whaling equipment when

---

*Umiat is the plural form of *umiaq*.

needed or cleaning and repairing the equipment they already have. In the northern communities, crew members must also survey and check the ice, searching out cracks and flaws, looking for weaknesses in the pack ice, and establishing a safe, solid route. Once camp is set up, lookouts are posted to scan for bowheads. While waiting for the whales to arrive, the crews may hunt seals or waterfowl. The captain explains to new crew members the way in which to hunt the whales and the parts of the body where the bowhead is most vulnerable.

The hunting of the whale is arduous and frequently fraught with danger. It involves not only pursuing a massive animal in a small boat and attempting to kill it with a relatively small harpoon gun—far smaller than anything used by modern whalers on their steel ships—but doing so in water still dotted with chunks of ice, both large and small, which could easily rip a hole in the light *umiaq* and cast its crew into freezing Arctic seas.

Sometimes the hunt may take hours and be unsuccessful. A whale may be struck with a harpoon but disappear from view, either to live on or to die slowly, out of sight. In the latter case, the whale is lost, even if it is recovered later. It takes little time for a whale carcass to heat up internally, spoiling the meat and organs and making them unsuitable for human consumption. Sometimes, however, everything falls into place and a hunt ends in success. When that happens, it is undoubtedly a testament to the skill of the hunters; to the Iñupiat, however, it goes deeper than that. The bowhead, they believe, is so intelligent and powerful that they could never kill one without its cooperation; a successful hunt is the result of a whale electing to give itself up to them. Accordingly, every aspect of the hunt is conducted in a particular way, with rituals followed as closely as possible, so as not to annoy or frighten the whales and to demonstrate requisite respect.

Following the kill, the whaling captain, his family, and his crew get first choice in whale meat and *muktuk*—the whale's skin and underlying blubber. The rest is then distributed throughout the community and shared with others: whaling towns that were unable to obtain their desired number of whales; communities that are unable, by location or for other reasons, to hunt bowheads; and individuals who have left the villages and are scattered throughout Alaska.

The hunt is marked also by celebrations that last throughout the year. After

each successful whaling season, the community joins together for *nalukataq*—
a coming together of people from all surrounding communities for a potlatch
(a large feast of traditional dishes), Eskimo dancing, and other festivities.

For generation upon generation, the hunt was conducted in much the same
way. Little was wasted: *muktuk,* meat, and organs were eaten; bones were used
to frame sod houses, or *iglus,* or used as runners for sleds; baleen was woven
into baskets; liver membranes were used to cover drums; oil was used for lighting and heating. The number of whales taken throughout the Arctic was in the
low teens each year.

The bowhead hunt was not only at the center of Iñupiat life; it defined it.
The cooperative hunting activities throughout the year and the communal patterns of sharing the whale integrated the society as a whole. The steady, gentle
rhythms of the Arctic year ebbed and flowed with the whales' migration, the
hunt's preparation, the hunt itself, and the subsequent celebrations. To any
hunter who chose to contemplate it, this must have seemed a rhythm that could
pulse the same way for eternity.

But it was not to be. One day in July 1848, just south of the Diomede
Islands, a strange object appeared about twenty-five miles off Cape Prince of
Wales, and seven *umiat* set off to investigate. It was a large ship, larger than any
the Iñupiat had seen before, but as the *umiat* approached, it seemed to turn and
move away. Out of sight of the natives, the ship became enveloped in a thick
fog; when the fog lifted a day later, the vessel was surrounded by bowhead
whales. The crew promptly lowered a boat into the water and killed one of
them. Examining their catch, the men marveled at the whale's thick blubber
and the tremendous length of its baleen. They wondered aloud at the possibil-
ities that lay ahead, the money that could be earned from what appeared to be
limitless numbers of whales in these uncharted waters.

They were the first commercial whalers to visit the western Arctic. They
would not be the last.

Pathways to the East

The journey that brought that first whaling ship to the Bering Strait began
more than three hundred years earlier on the other side of the world, as a result
of a quest for riches of an entirely different nature.

On May 11, 1553, three ships passed Greenwich, a borough of London, while sailing down the River Thames en route to the Arctic. Leaving England behind, they sailed for several weeks, traveling through the North Sea and along the coast of Norway until the fleet was dispersed by a raging storm, and two of the ships—the *Bona Aventura* and *Bona Esperenza*—were swept east into the Barents Sea. There, barely recovered from the storm's beating, they found themselves confronted with a sea of ice. Continuing any farther was plainly out of the question; but so, as the floes closed in around them, was returning the way they had come. They sought shelter on the Murman Coast of the Kola Peninsula, where they elected to spend the winter. By the time summer rolled around, extreme cold and scurvy had taken their toll: every crew member of both ships was dead.

Unaware of the plight of the other two ships, the third vessel, the *Edward Bonaventure,* rounded the North Cape and continued on to the White Sea. As his crew wintered in the town of Arkhangel'sk, the ship's captain, Richard Chancellor, sledded to Moscow, established ties with businessmen and authorities there, and opened up trade with Russia. Henceforth, the expedition's organizers—who at the outset of the voyage had styled themselves the "Mysterie and Company of Merchant Adventurers for the Discoverie of Regions, Dominions, Island and Places Unknown"—became known as the Muscovy Company. Queen Elizabeth I granted them a monopoly on trade with Russia and all areas to the north.

It was a fine prize, but the aims of the expedition had initially been fixed much farther to the east. In the lands of the Orient lay riches, from teas and spices to gold and silks, for any nation able to establish and maintain commerce with them, and in the mid-1500s that chalice was firmly in the grasp of the Spanish and Portuguese. The latter had established a solid grip on the routes and ports along Africa and to the east; the former, following Columbus' encounter with the Americas, had effectively seized control of what was known of the Western Hemisphere. The ways to the south, west, and east therefore effectively blocked, England had little choice but to look elsewhere. And so the English turned their gaze north, in the hope of finding a passage to either the northwest (i.e., across the top of North America) or the northeast (above Eurasia) that would lead straight to the riches of Cathay.

They were not alone. Shortly after their search began, the English were joined in their quest by the Dutch, newly independent of Spain and anxious to develop their own trade routes and sphere of influence. At the forefront of the Dutch effort was Willem Barents—probably, argues Arctic historian Richard Vaughan, "the greatest sixteenth-century Arctic explorer," a man who combined "outstanding navigational skills with excellent leadership, and had the courage and resolution necessary for the navigation of unknown ice-choked seas." In 1594, Barents' attempt to push north of Novaya Zemlya was denied by thick pack ice in the Kara Sea and a reticent crew; two years later, however, he tried again, as navigator on board a ship commanded by countryman Jacob van Heemskerck. Heemskerck, like Barents, was a fabled mariner; it has been written that he was "incapable of fatigue, of perplexity or of fear." Both men's qualities were to prove essential as their ship and crew encountered conditions and privations such as few European explorers had even imagined.

Barents and Heemskerck succeeded in rounding the northern tip of Novaya Zemlya and pushing on to the east. But although Barents had, notwithstanding his difficulties on the previous expedition, been brimming with confidence that this route would open up and yield a clear passage to the Orient, the ship was soon engulfed by encroaching pack ice. The hulls creaked and protested, threatening to split. The crew grew anxious; one of them, Gerrit de Veer, recorded later that the sight and sound of the ice enveloping the ship in its suffocating embrace "made all the hairs of our heads to rise upright with fear." In late August, Heemskerck, Barents, and crew sought refuge in a shelter they called Ice Haven, on the northeastern coast of the island. It proved inadequate; the ice continued to come at them from all directions, and the realization dawned that if they did not extricate themselves swiftly, they would be forced "in great cold, poverty, misery, and grief to stay all that winter."

Within a matter of weeks, they had resigned themselves to doing just that. Taking advantage of a good supply of driftwood that lay on the shore—and fortuitously so, "for there was none growing upon that land"—they built a shelter large enough for the whole crew and became the first Europeans to spend the winter on an uninhabited Arctic coast.

It was a punishing ordeal. Polar bears were a constant threat; more than one showed too much interest in the human interlopers and had to be shot. The cold and the dark weakened the crew both physically and psychologically. They attempted to protect themselves from the elements by plugging up every conceivable leak and crack in their dwelling, but the process proved so effective in keeping out fresh air that they "were taken with a sudden dizziness in our heads . . . and found ourselves to be very ill at ease," until the door was opened once again and "the cold, which before had been so great an enemy to us, was then the only relief we had."

Finally, the sun returned. With each passing day the sunlight lasted longer, and as summer approached, the crew began to make preparations for their escape. The ice had damaged the ship beyond repair, so they were compelled to use their two small boats for the hazardous journey home. Remarkably, they made it, threading nervously through the slowly retreating ice, rounding Novaya Zemlya, and hugging the western coast before eventually making landfall on the Kola Peninsula. In perhaps the least concise title ever bestowed on a book about polar exploration, Gerrit de Veer called his account, which proved wildly popular on publication, *The True and Perfect Description of Three Voyages, so strange and woonderfull, that the like hath neuer been heard of before: Done and performed three yeares, one after the other, by ships of Holland and Zeeland, on the North sides of Norway, Muscouia, and Tartaria, towards the Kingdomes of Cathia and China; shewing the discouerie of the Straightes of Weigates, Noua Zembla, and the countrie lying under 80 degrees; which is thought to be Greenland: where neuer any man had bin before: with the cruell Beares, and other Monsters of the Sea, and the unsupportable and extreame cold that is found in those places. And how in that last Voyage, the Shippe was so inclosed by the ice, that it was left there, whereby the men were forced to build a house in the cold and desart Countrie of Noua Zembla, wherein they continued 10 monthes togeather, and neuer saw nor heard of any man in, most great cold and extreame miserie; and how after that, to saue their liues, they were constrained to sayle aboue 350 Dutch miles, which is aboue 1000 English, in little open Boates, along and ouer the Maine Seas, in most great daunger, and with extreame labour, unspeakable troubles, and great hunger.*

Willem Barents did not live to enjoy the fame and praise that greeted his crew on their return home. Severely weakened during the winter, he died on the return journey, on an ice floe, just as the boats waited for a lead to open for them to round Novaya Zemlya on the journey home.

Meanwhile, the English, discouraged from their quest for the Northeast

Passage by the seemingly impenetrable ice encountered by one expedition after another, had been concentrating their efforts on a search for a route to the northwest. Between 1576 and 1587, they had launched six voyages in this direction, but when none of these proved successful, there was a pause of more than two decades before a brilliant young explorer named Henry Hudson set out on the journey on which he discovered the bay now named after him. Unfortunately, despite a sharp mind, Hudson was no great leader of men, and this voyage would be his last: on at least two occasions during the expedition, rumblings of mutiny pervaded the decks until finally he was cut adrift in a small boat, with a half-dozen companions, and left to die in the icy waters.

In 1607, three years before meeting this untimely end, Hudson had been dispatched by the Muscovy Company to test the veracity of a theory first advanced in 1527 by a Bristol merchant called Robert Thorne: that rather than northeast or northwest, the best path to success lay straight north, across the North Pole. Supporters of this view held that the cold and ice of the northern regions was at its greatest on the latitude of the Arctic Circle, and that farther north, any ice in the polar regions would melt during the long Arctic summer—that "the sun at the far north was rather a manufacturer of salt than of ice."

This, as we now know, is demonstrably incorrect; and so, Hudson did not find any Trans-Polar Passage. He did, however, discover Jan Mayen Island, in the Barents Sea, and push far north to the northern limits of a bleak, rocky archipelago. Heemskerck and Barents had seen that same archipelago during their voyage, named it Spitsbergen (from the Dutch *spits,* "pointed," and *bergen,* "hills"), assumed it was part of Greenland, and dismissed it as barren and uninviting. Hudson, however, returned home with a more favorable report—not of the islands themselves, of which his view concurred with that of his fellow explorers, but of the wildlife that surrounded them. The bays of both Spitsbergen and Jan Mayen, he recorded, were teeming with whales.

Up to this time, commercial whaling among Europeans had essentially been restricted to the Basques, a coastal people of uncertain origin who lived in a corner of the western Pyrenees, at the meeting point between France and Spain. Beginning about the eleventh century, they had systematically hunted whales—first in the Bay of Biscay and then, starting in the early 1500s, off

Labrador and Newfoundland—and exported the products, principally oil for heating and lighting, to much of Europe. The whales they hunted made relatively easy targets. They swam slowly and close to shore; their plentiful blubber not only surrendered a rich supply of oil but also ensured that they obligingly floated when killed, making them easier to retrieve. They were, all told, the "right" whales to hunt; and in unfortunate recognition of the nature of their first sustained contact with humans, right whales is how they are known today.

The whales that crowded around Hudson and company in the bays of Spitsbergen were not right whales. Although certainly similar, they were larger, an altogether different species, one thus far unfamiliar to the English—and, indeed, to virtually all Europeans of the seventeenth century. They were bowheads.

✷ ✷ ✷

The earliest account of bowhead hunting by people other than Eskimos comes from King Alfred of Wessex, who in the late ninth century recorded a voyage to the entrance of the White Sea by Ohthere, a Norseman in his service. Ohthere reported that people in the region hunted "horse-whales," which from the description appear to have been walruses. They also sometimes killed whales that he claimed were forty-eight or even fifty ells in length—an exaggeration for sure, as that translates to more than 180 feet, but almost certainly a description of bowheads.

It is also possible that some bowheads were hunted in Scandinavian and northern European waters several hundred years later; indeed, the Labrador hunting grounds of the Basques overlapped the likely southern limit of the bowheads' range, as well as the northerly reaches of the right whales, and it therefore seems safe to assume that bowheads constituted a certain percentage of their catch.

But Hudson's observations spurred the first organized commercial hunt in which bowheads were specifically targeted. In 1576, following accounts of plentiful sightings along the routes they plied, Queen Elizabeth had granted the Muscovy Company exclusive rights to hunt whales "in any seas whatsoever"; but despite frequently drawing up plans to do so, not until after Hudson's voyage was the company spurred into action.

In 1611, the Muscovy Company sent two ships, the *Mary Margaret* and the *Elizabeth,* to Spitsbergen to hunt whales, carrying with them six Basque harpooners. Having waited almost forty years to conduct their first whaling operation, the company could scarcely be accused of not having had time to prepare properly. Nonetheless, the voyage was a disaster: the crews killed more than a hundred walruses but just one whale before the *Mary Margaret* was wrecked in the ice. Its crew took to the boats only to become separated, after which two of the boats encountered the *Hopewell,* an "unauthorized" whaler out of Hull commanded by Captain Thomas Marmaduke. Marmaduke rescued the men and headed back toward the wreck of the *Mary Margaret* to try to secure the ship's catch for himself. Meanwhile, the other boats, including one carrying the *Mary Margaret*'s captain, Thomas Edge, had come across the *Elizabeth,* and they, too, returned to the *Mary Margaret.* There, to their surprise and displeasure, they found the *Hopewell.* To preempt the efforts of the Hull ship, Edge ordered the cargo of the *Elizabeth*—just walrus hides and blubber, the crew of the *Elizabeth* having been even less successful than that of the *Mary Margaret*—to be put ashore and the *Mary Margaret*'s catch transferred. But in the process, the *Elizabeth* became so light "that she upset and was lost." Now both of the Muscovy Company's ships were gone, and Edge had no option but to ask the *Hopewell* to take crew and cargo back to England. Adding insult to injury, Marmaduke forced them to pay for the service.

It could hardly have been a less auspicious start. Nonetheless, by 1613, the waters of Spitsbergen were choked with whaling ships. The bounty they sought was so plentiful that, according to some accounts, it was almost impossible to see the sea for all the whales that swam at the surface. But as far as the Muscovy Company was concerned, there were not enough to share. The company returned to the area with a seven-vessel fleet, brandishing a new royal charter once more granting them a monopoly, and with firepower to back it up. The *Tiger,* the fleet's well-armed flagship, ran off the foreign whaling ships, either confiscating their cargo and forcing them to leave the whaling grounds or allowing them to stay provided they surrendered a portion of their catch.

Initially, pickings were good. That 1613 season, just one of the company's fleet, the *Mathew,* had secured 184 tons of oil and 5,000 pieces of baleen within a matter of weeks. By 1617, the English were killing 200 whales and taking home more than 11,000 barrels of oil each year. But perhaps because they spent so much time chasing off rivals rather than actually whaling, the Muscovy Company posted a substantial loss in 1613, and the following year was an absolute disaster. The whales arrived in the area late in the season, the temperature remained cold, and ice clogged the fjords where most of the whaling took place. Most ominous of all for the long term, however, was that this time, the other whalers—steaming at their treatment the previous year at the hands of the *Tiger*—fought back.

In response to an edict from the new English king, James I, granting the Muscovy Company the authority to annex Spitsbergen for England, the Dutch formed the Noordsche Compagnie, or Nordic Company, giving its members sole right to hunt whales on all coasts from Novaya Zemlya to Davis Strait, between Canada and Greenland. To their fleet of eleven ships, several of which were well armed, they added three warships. Cowed by this show of force, the English wisely elected to sue for peace. The two countries agreed to divide up the western coast of Spitsbergen between them, the English securing the four harbors that had proved to be among the most productive thus far and the Dutch being free to make use of the rest of the coast. It was an uneasy truce, and it did not hold. In 1617, irritated by Dutch trespassing in their harbors, the English seized a Dutch ship, confiscated its cargo of whale oil and some of its equipment, and sent it packing.

The next year, the Dutch returned armed to the teeth. English admiral William Heley arranged a meeting with his Dutch counterpart, Captain Huybrecht Cornelisz, but the Dutchman, drunk, tried to attack Heley with a knife. The situation reached the boiling point when later in the season, Cornelisz summoned Heley and the other British captains and ordered them to hand over their catch. They refused, and the Dutch opened fire, blasting the English fleet with cannons; helped themselves to casks of oil and bundles of baleen; took food, liquor, and supplies; and left the battered English ships to patch themselves up and limp home.

Again, the two sides agreed to negotiate a peace, but whereas previous arrangements had failed, this was one the whalers themselves wanted. It was difficult and dangerous enough trying to kill whales in icy Arctic waters without being shot at by another country's warships. Besides, fighting and chasing other vessels cut into whaling time and, hence, profits. Once again, the English and Dutch agreed to divide Spitsbergen between them. The Muscovy Company retained the western coast, and the Noordsche Compagnie moved to the northwest of the archipelago and the coasts along the north.

The Dutch got the better of the deal. Whaling in their part of Spitsbergen proved more profitable, whereas the English, despite having ostensibly relieved themselves of their conflict with the Dutch, had to contend with continued infighting among different factions, all wanting a piece of the whaling pie. In 1625, the Muscovy Company fleet arrived at Spitsbergen to find that its stores and equipment had been destroyed by rivals from Hull and York. In the 1630s, prices for whale oil began to fall in England, and in the 1640s the country became embroiled in a civil war, causing its focus to be shifted away from the whaling grounds.

Other countries entered the fray around Spitsbergen—France, Denmark, Germany—but the Dutch remained dominant. Learning from the Basque experts whom they frequently employed on their ships, they conducted their whaling in much the same manner as the Biscayans had done: operating from longboats, they would launch one or two hand harpoons into each whale and then hang on to the quarry as it dived and tried to swim away. Other boats would join the chase, throwing more and more harpoons, wearing the whale down until finally it was exhausted. Then they would finish it off with a series of lance thrusts, tow the carcass to shore, and flense the blubber and remove the baleen.

During the middle years of the seventeenth century, Spitsbergen whaling grew and grew. In 1619, the Dutch began operating a quasi-permanent settlement on Amsterdam Island, in the extreme northwestern corner of the archipelago. Rejoicing in the name of Smeerenburg, or "Blubber Town," it consisted of a tryworks, warehouses, and even a church. When in 1642 the Noordsche Compagnie's monopoly lapsed, the field was thrown wide open, and hundreds of vessels took advantage. That year, 30 to 40 whaling ships were outfitted in the Netherlands; a little more than forty years later, the number had risen to 246. Writing in 1820, William Scoresby calculated that at its peak, the Dutch fleet transported 18,000 men to the area each year in pursuit of whales.

Those were the glory days of Dutch whaling, but, of course, they could not last. As Scoresby wrote: "The whales, which were so constantly and vigorously pursued, in a great measure left the bays, receded to the sea, and eventually to the ice." Put another way, within a matter of decades the bowhead whales around Spitsbergen were all but wiped out.

As a result, whalers were forced to head north—closer to, and even into the heart of, the pack ice. And they went west, closer and closer to Greenland, until eventually they were regularly whaling off Greenland's eastern coast. On one level, it was a successful move: between 1675 and 1721, the Dutch killed almost 33,000 whales. But much of that success came early in the endeavor. Around 1690, average annual catches were approximately 2,000 whales; by 1720, they had fallen to about 800. The remaining whales were to be found ever deeper in the ice, treacherous territory for wooden sailing ships. As a consequence, it was not just bowheads that perished in the frozen waters of the northern seas.

Between 1661 and 1718, an average of 149 Dutch whalers sailed each year into the Greenland Sea; about 1 in 25, or an average of 6 per year, did not return. In some years, particularly bad conditions made the losses even greater: in 1678, the Dutch lost 18 ships out of 110; in 1685, 23 out of 212.

In the same way exploration had led to discovery of the whaling grounds off Spitsbergen, the search for new whale populations spurred further geographic discoveries in the Arctic. As acclaimed historian Jeanette Mirsky observed, "discovery for discovery's sake ceased; it was carried on and kept alive only by whalers who combined it with their business." Now it was the whalers who covered great areas in search of new destinations, and the Dutch did so more than any others.

Sometime in the late seventeenth century, Dutch whalers rounded Cape Farewell, on the southern tip of Greenland, and gingerly nosed their way north up the island's western coast into an area known as Davis Strait. There, they came across a population of bowheads that had thus far escaped their attention. It was a whole new vein to mine, but the Dutch largely opted out of the opportunity. The ice in Davis Strait was even greater than that east of Greenland, and losses were mounting. After an initial show of interest, Dutch whalers concentrated almost exclusively on the eastern side and essentially left Davis Strait alone.

Their absence gave the British an opportunity to enter the fray once more. Driven by industrialization—in particular the burgeoning textile industry, which used whale oil for "fulling," cleaning wool prior to spinning it—the need for whale products was increasing again. To encourage British subjects to take full advantage of this new opportunity, the parliament in London announced in 1733 it would subsidize the industry to the tune of a bounty of twenty shillings per ton for every whaling ship over 200 tons. When that did not prove sufficient incentive, the bounty was raised to thirty and then, in 1749, forty shillings per ton. Thereafter, the British whaling industry experienced a rapid rebirth. In 1749, the year the bounty was upped to forty shillings, 2 ships were fitted out for the Greenland fishery. The following year, the figure rose to 20. Within a further six years, it was at 83, and by 1788 it had reached 253.

During England's long absence from commercial whaling, the Muscovy Company had largely disappeared from view, and the majority of British whalers came from Hull and Whitby in England and Dundee and Peterhead in Scotland. They set out for the Greenland whaling grounds in April, perhaps pausing at the Orkney or Shetland Islands on the way to pick up extra crew. The journey generally took roughly ten days, with the ships sailing north as far as about the fifty-eighth parallel and then turning west until south of Cape Farewell. As Greenland drew closer, temperatures began to plunge. Ocean spray froze on the rigging and rudders. Crew members piled on clothing and lit small fires in their cabins in a desperate attempt to keep warm. And then, out of the mist, loomed the icebergs, calved off from the giant Greenland ice shelf, silent assassins capable of bringing any vessel's voyage to a premature and deadly end. Finally, the ships rounded Cape Farewell, entered Davis Strait, and steeled themselves for battle with the immense fields of pack ice that came streaming down to meet them from the high reaches of the Arctic.

As with the Dutch off Spitsbergen, the British did not have Davis Strait entirely to themselves. In addition to the remaining Dutch, the whaling grounds were graced by the Germans, the Danish, and to a lesser extent the French and Spanish. Here, too, at least for a while, were Americans, although the extent of their involvement seems uncertain. Much of their activity in the region, it has been suggested, was curtailed by raids by privateers, and from 1750 to 1784, the Americans apparently avoided the region altogether. What-

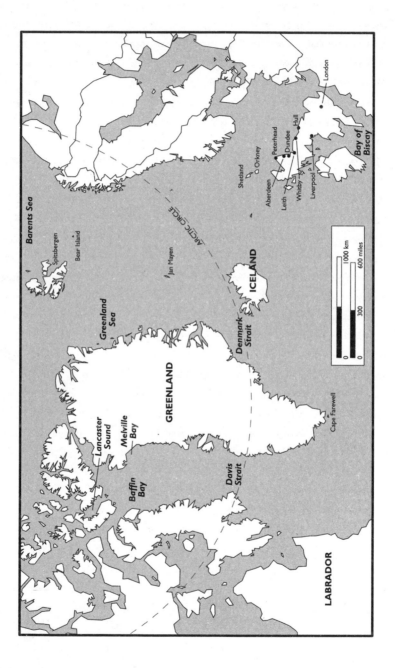

Barents Sea

Spitsbergen

Bear Island

Greenland Sea

ARCTIC CIRCLE

Jan Mayen

GREENLAND

Lancaster
Sound

Melville
Bay

Baffin
Bay

Davis
Strait

Denmark
Strait

ICELAND

Cape Farewell

LABRADOR

Shetland

Orkney

Aberdeen

Peterhead

Leith

Dundee

Whitby

Hull

Liverpool

London

Bay of
Biscay

1000 km

600 miles

0   300

0

ever the case, it seems likely that even as the British fought bowheads and ice fields in Davis Strait, American whalers were concentrating most of their efforts on sperm and humpback whales, which they hunted in both Atlantic and Pacific waters. By the middle of the nineteenth century, most American whalers probably still knew little, if anything, about the bowhead. At least one man, however, had been quietly making a study of what was at the time known only as the Greenland, or polar, whale.

New-Fangled Monsters

Thomas Roys was a man of tremendous will and fortitude who devoted his life to the whaling industry and spent years dreaming up new ways to catch and kill whales. Testing one such device—a harpoon propelled by a rocket from a shoulder launcher—he recoiled when it went off with a massive bang. He asked if anyone was hurt by the blast; there was no reply. "I then saw lying upon the deck a finger with a ring upon it which I knew, and looking I saw my left hand was gone to the wrist."

Some years previously, in 1844, he had been attempting to harpoon a right whale in the Pacific Ocean when the whale's thrashing flukes broke two of his ribs. It was to prove a serendipitous event because it led to two encounters that galvanized his interest in the bowhead. The first was with a Russian naval officer he met while recovering from his injuries in Petropavlovsk, on the Kamchatka Peninsula, who told him tales of strange whales he had encountered during a voyage north of the Bering Strait. The second, shortly after returning to his ship, was with a Danish captain, Thomas Sodring. Sodring and his crew had been whaling off Kamchatka when they caught some whales that at first blush appeared to be right whales, but were larger and had more blubber, to say nothing of the enormous lengths of whalebone hanging from their upper jaws. Convinced that both men had encountered the same whales that had been hunted in the Atlantic Arctic since at least the seventeenth century, Roys purchased Russian charts of the region, thumbed through the accounts of explorers who had sailed north of the Bering Strait, and quietly began to hatch a plan.

He returned home in May 1847; when he set out again two months later as captain of a bark called the *Superior,* he did so in the employ of the Grinnell Minturn Company with instructions to head to Kerguelen Island, on the fringes of the Antarctic. And so he did, but the voyage was unsuccessful, with the crew killing just three whales. Following a similarly unfruitful excursion to the South Pacific, the *Superior* refitted in Hobart, Tasmania, and here Roys made his move. He mailed a letter to his employers outlining his belief that there were much greater rewards to be found far to the north, in the Bering Strait, and informing them that was where he was taking the *Superior*.

Roys timed the mailing of his letter so that when the company learned of his intentions, he would be well on his way and unable to be reached. Nonetheless, he undertook the mission with an acute sense of what lay before him should he fail: "There is heavy responsibility resting upon the master who shall dare cruise different from the known grounds," he wrote later, "as it will not only be his death stroke if he does not succeed, but the whole of his officers and crew will unite to put him down." Indeed, so great was the anxiety of the crew at venturing into such icy, hostile waters that he afterward reflected, "I actually believe if they had any hope that open mutiny would have succeeded they would have tried it to get away from this sea."

The farther north they sailed, the more the crew's anxiety increased. It grew when those seven *umiat* rowed out toward them from Cape Prince of Wales, and it did not dissipate when, emerging from fog the following day, they found themselves in the company of whales. For after initially believing them to be humpbacks, they soon realized this was a species they had not seen before—a "new-fangled monster," they called it—and, accordingly, one they viewed with extreme caution.

Roys, however, knew from reading accounts of whaling in Greenland and Spitsbergen that the bowhead was a relatively docile beast, and sure enough, the first whale they pursued proved remarkably easy to strike with a harpoon. Immediately thereafter, however, it "went down and ran along the bottom for a full 50 minutes and I began to think that I was fast to something that breathed water instead of air and might remain down a week if he liked." But "he then came up and was immediately killed."

When the crew cut into the whale's carcass, they were astonished to discover that its baleen was fully twelve feet long, much longer than that of most whales. When they rendered its blubber, it produced 120 barrels of oil—

almost as much as they had secured during their entire trip to Kerguelen Island.

Suitably inspired, Roys pushed yet farther north, 250 more miles all told, spending about a month north of the Bering Strait and killing eleven whales, which yielded a total of 1,600 barrels of oil. On August 27, 1848, weighed down with its blubbery bounty, the *Superior* passed south of the strait and headed for Hawaii; from the moment it put in at Honolulu, news of its plunder spread like wildfire.

The following year, fifty whaling ships set out for the Bering Strait and caught more than 500 bowheads; in 1850, the number of bowheads killed swelled to more than 2,000. In 1852, the waters of the strait were crisscrossed by the paths of more than 220 whaling ships—more vessels, according to historian John Bockstoce, than have been in the area in any year since. More than one-quarter of all the bowheads that would eventually be taken in the western Arctic were killed during a few years. Not surprisingly, almost as soon as it had begun, bowhead whaling in the Bering Strait all but collapsed. In 1853, the total catch was only about one-third what it had been the year before. By 1854, it had plunged to just more than 100 whales, and the following year, the whalers abandoned the Bering Strait completely.

For a few years they turned south, switching their attention to the Sea of Okhotsk. Bowhead whaling there had begun in 1845, when the Danish whaler *Neptun* became the first "foreign" whaler to take a bowhead in the Pacific; by 1847, thirty ships were plowing the Okhotsk's waters. Within twenty years, an estimated 18,000 whales had been killed, and the catches had declined to almost zero.

With nowhere else to go, the whalers returned to the Bering Strait region. But the apparent absence of bowheads that had caused the fleet to abandon the strait had not ended during their brief sabbatical. If they were to continue hunting whales in the western Arctic, the whalers were going to have to push even farther north, into the treacherous ice-strewn waters of the Chukchi and Beaufort Seas, where, they reckoned, the rest of the bowheads could be found.

We have seen already what happened to the Dutch when they were obliged to adopt the same strategy after wiping out the bowheads around Spitsber-

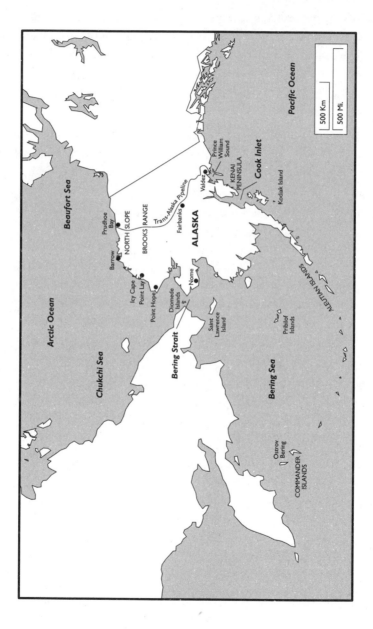

gen—although catches had increased, so had the number of ships lost in the ice. West of Greenland, the British had followed a similar path, forcing their way through the pack ice of Davis Strait and into the open water of Baffin Bay, where the last remaining bowheads had found sanctuary behind the barrier of ice. Such a move enabled British whalers to prolong the life of the Greenland fishery, but it came at a terrible cost: between 1819 and 1843, a total of 82 ships were lost in Davis Strait. The worst season of all was in 1830, when 19 of 91 ships perished, most of them in Melville Bay, on the northwestern corner of Greenland, as they waited for the ice to move off enough for them to push west. At one stage, more than a thousand crewmen were adrift on the ice, living in tents or beneath whaleboats, watching as their ships sank beneath the ice and praying another would provide them passage home. Of the 72 ships that survived the season, 12 were seriously damaged and 21 returned home "clean"—that is, they did not manage to kill a single whale. A few years later, in 1835 and 1836, the fleet suffered additional back-to-back disasters, losing yet more ships and a great many sailors and leaving the Davis Strait whaling fleet reeling.

As with the British and the Dutch in the eastern Arctic, so also with the Americans in the west. In the decade of the 1870s, 58 whaling ships were lost in the icy waters of the Chukchi and Beaufort Seas—32 of them in 1871, when capricious winds pushed thick pack ice toward the coast, trapping and overrunning the fleet, crushing ships and contemptuously forcing them onto the shore.

The disasters of 1830, 1835, and 1836 that struck the Davis Strait whalers knocked the wind out of the British Greenland whaling industry. Although bowhead whaling in the eastern Arctic would continue until the early twentieth century, the cost of losing so many ships, as well as the decline in the number of whales, made the exercise decreasingly profitable. By 1841, only about one-fifth as many British whalers visited Davis Strait as before.

The calamitous events of 1871—combined with subsequent losses of ships in the ice throughout the 1870s—seemed set to do the same to the American fleet in the western Arctic. The whales were becoming scarcer, the remaining whaling grounds more dangerous. Then, beginning in the late 1850s, the emergent petroleum industry kicked the legs from under the market for whale oil.

In 1876, only 20 whaling ships made the journey to the Beaufort and Chukchi
Seas, and 12 of them were lost in the ice near Point Barrow.

In the face of all these losses, it seems the only reason the ships' owners did
not abandon the business completely was that profits on whaling were by now
so marginal that they found it difficult to sell the ships. Accordingly, they
elected to continue running the fleet at a small profit, perhaps awaiting one
more opportunity to hit pay dirt. Even if that never quite happened, good fortune did enable the industry to find one final lease on life, which would last
another thirty years or so. The saving grace was the fact that in response to the
growing fashion for whalebone corsets and hoopskirts, the price of baleen had
gone through the roof—and no species of whale had greater supplies of baleen
hanging from its upper jaw than the bowhead. Around the same time, the
advent of steam-powered ships enabled the industry to force its way through
the ice and into the final, darkest corners of the Arctic, where the last bowheads had so far avoided the whalers.

In 1884, following the lead of the Iñupiat, commercial whalers established
a shore station at Point Barrow to take the bowheads during their spring and
fall migrations. Three years later, another shore station was set up at Point
Hope; soon, others were scattered along the northwestern coast of Alaska.
Meanwhile, steamships pushed east of Point Barrow. By now, the whalers knew
for certain that the whales they were hunting in the western Arctic were the
same species as those pursued off Spitsbergen and Greenland. They also
assumed—as had so many for so long before them—that there was a Northwest Passage, a clear stretch of water that allowed the whales to swim from one
part of the Arctic to the other. When, in 1887, a group of Iñupiat from Point
Barrow told Charlie Brower, a pioneer whaler and trader, that they had seen a
great many whales in Mackenzie Bay, far to the east in the western Canadian
Arctic, he dispatched one of his men, Little Joe Tuckfield, to investigate. Tuckfield set off in July 1888; after spending the winter in the eastern Beaufort Sea,
he returned, in August 1889, with the news that bowheads in the area were
"thick as bees." Furthermore, he continued, he had found a perfect natural harbor in Herschel Island. It was here that commercial bowhead whaling entered
its final phase.

For a little more than a decade, ships used Herschel Island as the base for their operations in the eastern Beaufort. Because of the length of the voyage from the home port—normally San Francisco—to the Mackenzie River delta region, it was impractical to sail or steam to the north, kill whales, and then come home: far too much time would have been spent in transit, with very little left for actual whaling. Consequently, they overwintered, whaling during spring and summer en route to Herschel; spending the winter in the island's freezing, windy shelter; and then resuming whaling the following season.

The arrangement allowed Beaufort Sea whaling to limp on for a few more decades, until by 1914 diminishing returns and increasing shipwrecks in the ice once more reduced the industry's attractiveness to investors, this time for good. This final phase of bowhead whaling also had one other, unintended consequence: the almost year-round presence at Herschel Island, and at Point Barrow and other shore stations along the Alaska coast, brought the whalers of the western Arctic for the first time into continuous contact with the region's natives.

The Great Dying

It would be easy to overstate the case, to paint the interactions between whalers and Eskimos as being uniformly negative, with the natives always on the receiving end. In fact, in both the eastern and western Arctic, many natives established special relationships with the whalers and went out of their way to help them. A young Inuk* from Baffin Island called Eenoolooapik showed British and American whalers open water full of bowheads in Baffin Bay to the west of Davis Strait. Several Iñupiat villagers appointed themselves conduits for trading between their compadres and Yankee whalers. Many others crewed with and worked for the whaling fleets, either on the vessels or at shore stations.

Interactions between whalers and natives were frequently benign and uneventful. In a book of Inuit memories of commercial whaling in the eastern Canadian Arctic, *When the Whalers Were Up North,* Dorothy Harley Eber quotes Joe Curley, an Inuk from Hudson Bay, as stating:

Inuk is the singular form of *Inuit.*

I really can't recall any disagreements between the Inuit and the [whalers]. Sometimes the whalers would have the mate do something he didn't want to do, but there were never feelings of inferiority or superiority between whalers and Inuit. They would help each other out as best they could. It must have been a long winter for those people to put in on their ships. Most of their arguments would occur because of the long winters; they longed for home. They were pretty good among the Inuit people.

From the whalers' perspective, too, many contacts were nothing but friendly. John Bockstoce—whose history of whaling in the western Arctic set a standard to which others can only aspire—quotes an entry from the journal of Mary Brewster, wife of the captain of the *Tiger,* out of Stonington, Connecticut, on an encounter in the Bering Strait in June 1849:

Plenty of native company. We have had company from the whole coast I believe: a chief by the name of Notochen with his wife and child who is the greatest man of the nation, I expect. We were caressed, touched noses together and as near as we can understand, they are to be our friends. . . . I suppose they had never seen a white female before. They brought presents of their garments and walrus teeth for which we paid them with tobacco. They are all smokers and chewers—even the children—and are extravagantly fond of it. . . . They stopped nearly all day. We gave them bread, combs, needles, thread and knives, and at 4 they left well pleased with their visit.

But even such innocuous encounters would, in the long term, have serious disruptive consequences. From the time they first appeared in the Arctic and even more so after they began to spend winters there—from Hudson Bay and Cumberland Sound in the east to Herschel Island and Point Barrow in the west—whalers looked to trade with natives for goods, such as food and furs, that would provide an extra degree of comfort in the harsh, hostile north, as well as for baleen that could be added to their cargo.

In return, they offered needles, knives, blubber spades, and other imple-
ments—tools to make the life of subsistence sealers, whalers, and fishers that
much easier. But over time, some native communities increasingly diverted
precious time and energy previously spent on sealing, whaling, and fishing to
the making of wares, such as clothing, that they could exchange for such tools.

The whalers also provided firearms, to which many hunters took with rel-
ish. The new weapons enabled the hunters to kill more caribou, reindeer, or
musk oxen—in some cases, to provide meat to trade to the whalers in return
for more firearms and other items. Others found themselves falling back on
hunting to compensate for dwindling numbers of whales. Whatever the motivation, newly equipped with rifles, they sometimes killed so many animals that
populations became depleted. In an ever decreasing spiral, this spawned an
increasing reliance on food and produce from the whalers on the part of the
natives. It forced them to spend yet more time producing goods for trade or to
become involved in the wage economy, working for the whaling fleets and, in
the process, contributing to the overexploitation of the resource on which they
had previously depended.

And some of the whalers provided alcohol.

In 1859, one whaling captain noted that he could "buy a hundred dollars'
worth of bone or fur for a gallon of rum which costs 40 cents at home." Some
Yankee whalers showed the Iñupiat how to distill alcohol from potatoes and
molasses, which they traded for slippers and sealskin tobacco pouches. Historian Dorothy Jean Ray observed that Eskimos "drank excessively from the first
drop of liquor," using it as a stimulant and an escape. It is an observation that
many whalers made for themselves: Eskimos, wrote one, "will take nothing
else if they can get rum and you can get anything they have got for a drink of
rum. Two good drinks will get them drunk and then they are happy." Mary
Brewster noted in August 1849 that the natives' "demand [was] usually for rum"
and saw the writing on the wall: "I think that if the whaling continues good
here they will not be the happy people they are now. They will naturally learn
many ways which will not be for their good. Liquor will only cause them to
quarrel. . . . I wish we had more temperance ships and more temperate com-
manders so the influence could be better."

To be fair, many of the whaling companies and captains strove to ban trading in spirits and to limit the amount of alcohol carried on board ship—for the sake of their crews as well as the natives. But enough did otherwise for Mary Brewster's observations to prove sadly prescient. On several occasions, aggressive Iñupiat, fortified by alcohol and in search of more, launched raids on whaling ships; in one notable instance, confrontation came to a head with the drawing of knives, the firing of guns, and thirteen dead Eskimos. During the winter of 1871, when more than thirty ships were trapped in the ice off Alaska, many Iñupiat became sick after rummaging through the ships for alcohol and, finding none, imbibing the contents of the medicine chests. Some of the whalers who overwintered at Herschel Island used the lure of whiskey to buy the sexual favors of native women. A few, derisively referred to as "squaw men" by fellow crewmen, set up camp on the ice with mistresses while the women's husbands sat at home supping the spirits for which their services had been sold; others enjoyed female company on a much shorter-term basis.

In the late summer and fall of 1878, whalers traded "barrels and barrels" of whiskey along the coastal villages of St. Lawrence Island. Writer David Boeri, who spent several seasons in the St. Lawrence Island villages of Gambell and Savoonga, recorded that following the trades, "islanders drank to the point of oblivion. They gave no thought to hunting, and by the time they recovered, it was too late: bad winds were upon them, and in their frenzy for whiskey, many Eskimos had traded away the very nets with which they might have caught enough seals to tide them over until hunting conditions improved."

In addition to the effects of winds and weather and the alcohol-induced state of the islanders, the area's walrus population had by then been massively depleted, making it still harder to find the numbers the villagers needed. Whalers turned to walruses whenever whales were hard to come by, and the slaughter of that species in the Arctic was as great as the massacre of the bowhead. John Bockstoce and Daniel Botkin calculated that the total catch of walruses in the western Arctic was close to 150,000; David Boeri suggested that between 1860 and 1880 alone, "ships' riflemen may have gunned down 200,000 walrus, a staggering number from a herd that harpoon-carrying Eskimos were becoming increasingly dependent upon for food in the wake of the whale herd's decimation." Some estimates place the total number of walruses killed throughout the western and eastern Arctic at an almost incomprehensible 4 million.

On St. Lawrence Island, the net result of the combination of whiskey, weather, and the cumulative effect of two decades of overhunting was disaster. During the winter of 1878–1879, the island was struck by famine; fully two-thirds of the population, by some estimates, perished, and six villages simply died out completely. St. Lawrence Island was not alone: severe starvation was reported the following winter at settlements along the Chukotka Peninsula and at Point Hope at various times in the 1870s and 1880s. In 1890–1891, famine reduced the population of King Island, in the northern Bering Sea, by two-thirds.

Nor was famine the only cause of huge numbers of deaths among Arctic natives as a result of the whalers' presence. The whalers, sapped by poor nutrition, too much whiskey, and the dryness of the polar atmosphere, succumbed to a range of illnesses—influenza, measles, typhus, and others—and these they passed along to the native populations, which had never before encountered such diseases and had little or no resistance to them. In 1840, during the first winter spent by a British whaler off Cumberland Sound, cholera claimed one-third of the local Inuit population. Fifty years later, on Southampton Island, in northern Hudson Bay, virtually the entire population was wiped out by typhus that originated in a local whaling station. In Alaska, it culminated in what is known in the Yup'ik language as *yuut tuqurpallratni*—"when a great many died," or the Great Death—which claimed the lives of more than 60 percent of Eskimos and Athabascan Indians in the first few years of the twentieth century. So great was the trauma felt by the survivors of this Great Death that bewildered and leaderless, they were easy prey for Christian missionaries, who convinced them that their fate was punishment for their godless existence. The missionaries bullied the Yup'ik and Iñupiat into suppressing their long-held spiritual beliefs and abandoning traditional practices such as Eskimo dancing, which the missionaries feared to be bordering on the satanic. Even speaking their own tongues was punishable. Confused and lost, many Iñupiat and Yup'ik communities became removed from the spiritual ways of being that had been their lifeblood just a generation before; only recently have they begun to recover and reassert their heritage.

Many of the early missionaries in Arctic Alaska followed in the whalers' wake, inspired particularly to right the wrongs done to the souls of Iñupiat, Inuvialuit, and white men alike by the reported debauchery at Herschel Island and other shore stations. But their presence in the region considerably outlasted the whaling fleets; by the time of the Great Death, the commercial whalers of the Bering and Beaufort Seas had begun packing their bags and heading home. The Pacific Steam Whaling Company, the principal whaling company during the latter years of the western Arctic fishery, closed its shore station at Point Barrow in 1896. Early in the twentieth century, the price of baleen collapsed, and by the outbreak of World War I, commercial bowhead whaling, in both the eastern and western Arctic, was over.

The Final Reckoning

John Bockstoce and Daniel Botkin estimated that the western Arctic bowhead fishery killed 18,650 whales between 1849 and 1914. Other researchers calculated that between the years 1660 and 1912, a total of almost 23,000 voyages to Spitsbergen killed more than 90,000 bowheads. In Davis Strait, more than 6,400 voyages between 1719 and 1916 yielded 28,695 recorded bowheads. Including a shorter-lived hunt focused exclusively on Hudson Bay in the nineteenth century, at least 120,000 bowheads were killed in the Atlantic Arctic by commercial whalers over the span of two and a half centuries. But this is at best a bare minimum. It does not include whales that escaped but were mortally wounded. Nor does it include whales killed off Spitsbergen by the Dutch and English prior to 1660, any of the whales killed in the seas east of Greenland, any that might have been taken as part of the Basque fishery off Labrador, or those killed by less significant whaling countries such as France and Denmark. Consequently, the total is certainly higher, perhaps substantially so.

The consequences of such hunting? The bowheads of the Bering, Beaufort, and Chukchi Seas probably numbered more than 20,000 and close to 30,000 before the arrival in the Bering Strait of Thomas Roys. The Scientific Committee of the International Whaling Commission (IWC) has estimated the present level for this population to be around 7,500. Larger to begin with than bowhead populations elsewhere and subjected to commercial exploitation for a shorter period of time, it is, comparatively speaking, a picture of health.

According to a 1993 review by the Society for Marine Mammalogy, the minimum pre-exploitation size of the Davis Strait and Hudson Bay bowhead population was estimated at 11,800. That same review estimated the total present population to be about 450. In Spitsbergen prior to the first voyage of the Muscovy Company in 1611, there were an estimated 24,000 bowheads. In recent decades, sightings of bowheads in the region have been sporadic at best. Optimists have suggested that the Spitsbergen bowhead population may now number "in the tens." It is, to all intents and purposes, extinct.

Commercial bowhead whaling in the Arctic has also left its mark on subsistence whaling. Inuit hunters in western Greenland are believed to have caught, on average, ten bowheads per year prior to the arrival of European whalers in Davis Strait. During an exceptional year, that figure could sometimes reach twenty or more. But as the Davis Strait bowhead population declined, the Inuit hunt dwindled to about one whale per year before ending altogether in the early nineteenth century. It was a similar story in eastern Greenland, with bowhead whaling all but finished by the early 1800s and the very last known landing of a bowhead taking place in 1880.

On the Chukotka Peninsula, where perhaps ten to fifteen whales were killed annually prior to the invasion of the commercial whalers, catches by native peoples apparently actually increased between the 1860s and the 1880s in response to the introduction of Yankee whaling gear and a market for baleen. But with the decline in the population and the cultural disruption resulting from the whalers' presence, whaling petered out here, too, with the last whale apparently killed in 1972.

There have been some attempts to revive native bowhead whaling. In early 2000, the last remaining Chukotka whaler, now nearly blind, visited Alaska for an operation to restore his eyesight so that he might be better able to pass on his hunting knowledge to a new generation. In present-day Nunavut, in the eastern Canadian Arctic, natives have since 1996 been granted—by the Canadian government, not by the IWC, of which Canada is not a member—a quota of one bowhead every two years. But only in Alaska has bowhead whaling continued largely uninterrupted into the twenty-first century—and even there, not without a fight and only after the story took a final, somewhat ironic twist.

Life for the Iñupiat would never be quite the same, even after the last

commercial whaler had left. They lived, some of them, in larger, more permanent communities than had previously been the case. Their traditional teachings
and beliefs came under assault from schoolteachers and missionaries. But they
retained their connection to the ocean and the sea ice and, especially, to the
bowhead whale. In the Beaufort Sea, at least, unlike the situation in Davis Strait
or eastern Greenland, there remained sufficient whales to sustain the Eskimos'
hunt; and so, for around a half-century after the commercial whalers' departure, they continued as they had before, killing perhaps twenty or so bowheads
each year. There were a few differences, some side benefits derived from the
Americans' presence: many Iñupiat now employed darting guns instead of hand
harpoons, for example. They mostly continued to use *umiat* rather than motor-
boats, though; the latter, they believed, scared off the whales. But such changes
were superficial; in essence, the Iñupiat hunt remained essentially the same as
it had for hundreds of years—and perhaps for thousands of years, back to the
time of the Old Bering Sea culture and beyond.

But then, in the early 1970s, another outside influence came to bear on the
Iñupiat. Massive oil fields had been discovered in Prudhoe Bay, off Alaska's
North Slope, and money began flooding into the region as companies sought to
extract the oil and transport it to the southern coast for export to the continental United States. It suddenly became easier for an Iñupiaq* to put together
enough money to purchase the necessary equipment and hire a crew to go
whaling. Crews and catches grew rapidly; by 1976, they had increased three-
fold. That year, Alaskan Eskimos landed a recorded 48 bowheads.

The increase in crews did not translate into a concomitant growth in
expertise. With a greater number of inexperienced men shooting at bowheads,
the number of whales struck (and presumably killed) but not retrieved grew,
in seeming disproportion even to the increase in whales actually landed. For
the 48 whales killed and landed in 1976, for example, an additional 43 were
known to have been struck and lost. Although the following year the landed
total fell to 28, the "struck and lost" figure soared to 77.

Such figures attracted the attention of scientists and environmentalists in

---

Iñupiaq is the singular form of *Iñupiat*. It is also an adjective and the name of the Iñupiat
language.

the outside world. When bowhead whales had been conferred protection from whaling worldwide in 1931, an exemption had been granted to native hunters. Now, however, that position was coming under increasing scrutiny.

In 1976, the IWC passed a resolution asking the U.S. government to reduce the catch and the rate at which whales were struck and lost. The next year, the National Marine Fisheries Service of the United States estimated that no more than 1,800, and perhaps as few as 600, bowheads remained in the western Arctic. At that rate, the Iñupiat hunt could finish off the population in a matter of years. Hearing that news, the IWC decided to act and rescinded the Iñupiat's exemption from the bowhead whaling ban.

Among the Iñupiat, the move was met with equal measures of shock and anger. They had *always* hunted bowheads. Why, all of a sudden, were people from thousands of miles away—most of whom had never been to Alaska, never met an Eskimo, never even seen a bowhead—telling them they could not do what their parents had done, their grandparents had done, and countless generations had done before them? They had never heard of this IWC or its concern about their bowhead hunt until it told the U.S. government, and the government in turn told them, that the hunt had to end. The concern of both scientists and environmentalists was genuine, but the Eskimos' fury was matched only by their bewilderment. White men had arrived uninvited in Iñupiat territory, had driven the bowhead out of the Bering Strait, and had almost exterminated it in the Beaufort Sea; now they were so concerned about the bowhead's survival that they were telling the Iñupiat they could not go whaling anymore.

Added to the Eskimos' sense of disbelief was the fact that there were, according to their observations, far more bowheads than the scientists claimed. The researchers counted only those whales they could see—those that rose to the surface while swimming through open-water leads. Most of the bowheads, the Iñupiat argued, spent long periods beneath the ice and thus escaped the scientists' attention.

Some within the native communities urged that the Eskimos ignore the ban and continue exercising their birthright, no matter the legal consequences. That stance lost out to those who elected to organize themselves politically and fight the white men at their own game. They formed the Alaska Eskimo Whaling Commission and began attending IWC meetings. They persuaded the commission to allow continued quotas for their hunting—although at first, the quotas granted (a total of twenty whales struck per year, likely to translate to just ten or so landed) were insufficient for their needs.

They also hired scientists to investigate their claims that the bowhead populations had been greatly underestimated. Some of these researchers, instead of attempting to count the whales visually, tested the Eskimos' theory by lowering hydrophones—specially made microphones that can record sounds underwater—through the sea ice. Sure enough, the equipment picked up the low-frequency moans of a good many bowheads—bowheads that had not been seen by those counting whales through binoculars. Over time, estimates and quotas went up.

The Iñupiat have also been working to improve whaling efficiency. Now, as much as 75 percent of the bowheads killed are retrieved and landed. Although some environmentalists continue to regard the hunt with a degree of suspicion, other authorities have called it the best-managed whale hunt in the world.

In 1998, the IWC's scientists agreed that the western Arctic bowhead population could probably sustain a catch of as many as 102 whales per year, although the actual quota set by the commission is lower: a maximum of 280 landed whales for the five years from 1998 to 2002. Meanwhile, studies suggest that the bowhead population is recovering from the overexploitation visited on it by commercial whalers and is growing by more than 2 percent annually.

As if to underline the bowhead's steady return, in 1999 the villagers of Little Diomede Island set out in an *umiaq* in pursuit of a whale migrating through open leads to the north. It was just a matter of miles from the island where Thomas Roys and the crew of the *Superior* killed their first whale in the western Arctic, and where only a few years later the bowhead had seemingly been driven from the Bering Strait entirely. Even as their relatives to the north had been able to sustain their whaling despite the effects of the commercial whaling fleet, the whalers of Little Diomede had found whales harder and harder to come by, and ultimately the hunt had all but ground to a halt. The villagers had not killed a bowhead since 1937; periodically, one would be spotted and a boat launched, but to no avail.

On this occasion, however, the chase ended in success. A darting gun was fired, and then another. The weapons found their target. The water boiled as the whale thrashed. Eventually there was silence, and the sea was still.

Chapter   3

Terra Incognita

Alexander Dalrymple glowers out from the canvas, his small mouth pursed into something a little less than a scowl. To the bottom of the frame, his large body begins a long, comfortable journey outward and downward, spreading unseen into what is presumably a very generous waistline. He could perhaps be handsome, after a fashion. But his features are spoiled by the extra weight they carry and by the tension that contorts them. He gives every impression of being a wealthy, perhaps self-made, man who regards himself highly and others with impatience and even contempt.

Historians have rarely been kind to Dalrymple. He has been described as "an obdurate, cantankerous Scot, of some ability, much conceit, and no sense of self-proportion." The great polar chronicler Hugh Robert Mill, writing in 1905, called him "fiery-tempered" and "too often ungracious and ill-natured." Furthermore, Mill wrote, "he expressed his views without reserve in a torrent of language which can only be described by quotation." Another account refers to him as "a Scot with a grievance, and one who will nurse it to keep it warm."

But as historian J. C. Beaglehole observed, "No contemporary would deny that Dalrymple was a passionate man." In addition to his undoubted vanity, wrote Beaglehole, few "would deny his ability, his knowledge, his enthusiasm, assiduity." In particular, Dalrymple was enthusiastic and passionate—and, to his mind at least, highly knowledgeable—about a subject that in the late 1760s and early 1770s was something of a Holy Grail for explorers, mariners, and merchants in much of Europe. Alexander Dalrymple was convinced of the existence of an enormous, hitherto undiscovered continent surrounding the South Pole, and he devoted years of persuasion and bile to convincing others of the importance of discovering it, exploring it, and claiming it for Britain.

The landmass Dalrymple envisaged was truly enormous, encompassing much of the Southern Hemisphere. It was so vast that in the latitude of 40°S it stretched for more than 5,323 miles, "a greater extent than the whole civilised part of Asia, from Turkey, to the eastern extremity of China." In support of this hypothesis, Dalyrymple cited a series of geographic discoveries in the Pacific Ocean—an Easter Island here, a coastline of Australia and New Guinea there—stitching them imaginatively together in one pan-hemispheric quilt. He stated categorically, "It rests to shew, from the nature of the winds in the South Pacific Ocean, that there *must* be a *Continent* on the *south*."

As much on deduction from past discoveries, Dalrymple's claims were based on the "analogy of nature." In his eyes, the fact that there were large landmasses in the Northern Hemisphere seemed to demand, even require, the existence of a landmass of similar size in the Southern Hemisphere.

In his belief in the existence of a huge southern continent, Dalrymple was not alone. Nor was he by any means the first. During the middle years of the eighteenth century, however, he was one of the most vocal—and ultimately most influential—advocates of a view at least two thousand years old.

"The Southernmost Man in the World"

In 322 B.C.E., Aristotle noted that constellations changed their apparent positions in the sky as a traveler moved north and south, and he observed that the shadow cast by Earth on the moon during a lunar eclipse was curved. These were, he offered, proof that Earth was a globe and not a flat disk. Demonstrating the ancient Greeks' love of symmetry and presaging Dalrymple's "analogy of nature," he contended that given the existence of a northern landmass surrounding the Mediterranean Sea, there had to be a counterbalancing landmass in the south. And just as the known world lay under the constellation of Arktos, the bear, so must the southern lands be under the opposite: Antarktikos.

In the first century C.E., the Roman geographer Pomponius Mela, borrowing from the works of, among others, Parmenides—a Greek philosopher writing in about 450 B.C.E.—proposed that the world was divided into five climatic zones: a frigid zone in the north, a temperate zone that included the Mediterranean region and neighboring lands, a torrid zone, another temperate zone, and finally a southern frigid zone. About a hundred years later, Ptolemy countered that the southern zone was fertile and populous—although he endorsed the general concept of a symmetrical Earth and agreed that the southern zone was divided from the habitable lands to the north by a region of fire.

But in this view lay the seeds of discord. In the centuries following Ptolemy, Christianity took root across Europe and the Church grew increasingly powerful and influential. Wherever it went, the Church proclaimed itself the ultimate arbiter of knowledge and set about suppressing any teachings that sat awkwardly with the absolute truths of the biblical canon. Ptolemy's view of Earth, and indeed the very concept of a southern continent, soon ran afoul of this burgeoning theological thought police.

For if there was indeed such a southern continent, if it was in fact inhabited, and if that inhabited landmass was, as suggested, separated from the rest of the world by an impenetrable zone of fire, that raised serious theological problems. Any people who lived there could not be descendants of Adam, a contention considerably at variance with the Scriptures, which made no reference to God creating Man anywhere other than in the Garden of Eden. Equally troubling was the fact that the Apostles had been commanded to go "into all the world, and preach the gospel to every creature." This southern land was said to be unreachable; the Bible did not record the Apostles having visited it; therefore, they could not have gone there; therefore, it could not exist. "Belief in Antipodes"—*anti-podes* being, literally, a place where people's feet were opposite—became a heretical offense. In 741, Pope Zacharias excommunicated an Irish priest, Virgil, for daring to articulate such a belief because "it would be admitting the existence of souls who shared neither the sin of Adam nor the redemption of Christ."

Criticism gave way to outright mockery. "Can anyone be so foolish," clucked Lactantius, tutor to the son of Constantine, the first Christian emperor of Rome, "as to believe that there are men whose feet are higher than their

heads, or places where things may be hanging downwards, trees growing back-
wards, or rain falling upwards? Where is the marvel of the hanging gardens of
Babylon if we are to allow of a hanging world at the Antipodes?" Disbelief in
human life on a southern landmass led to dismissal of that land's very existence.
And if that meant casting out the whole Greek ideal of a spherical Earth with
symmetrical landmasses counterbalancing each other, then so be it. Long
before it had been seen by human eyes, the idea of a southern continent had
become so threatening that it ultimately prompted the Western world to turn
its back entirely on the discoveries of Aristotle and Eratosthenes and to declare
that Earth was not a sphere at all, but flat.

"The Antarctic problem," wrote Hugh Robert Mill, "after being stigmatized
as heresy, had been crushed out of existence."

Meanwhile, however, some of the writings that had been spurned and, as
required, burned by the Christians were adopted and nurtured by Islam. In
particular, Muslim scholars embraced the writings of Ptolemy, and when in the
1400s Europeans embarked on a new wave of exploration and conquest,
Ptolemy's works—translated from Greek to Arabic and thence to Latin—were
waiting to help, inspire, and guide them.

There began a succession of voyages from England, Spain, France, Holland,
Italy, and Portugal. Over the next several centuries, these journeys gradually
peeled away the mysteries of the planet's geography, steadily reestablished the
veracity of a spherical Earth, and supplanted Ptolemy's vision of the world with
a progressively more accurate rendering.

Cape Bojador, south of the Canary Islands on the western coast of North
Africa, was long considered by superstitious sailors to be the farthest south it
was possible to sail. In 1433, Gil Eannes, squire of Prince Henry, passed this
invisible boundary for the first time. In 1486, Bartolomeu Dias sailed all the
way along the African coast, reaching the Cape of Good Hope. Ten years later,
his route was followed by Vasco da Gama, who sailed around the cape and into
the Indian Ocean. There was no great southern land here; if it existed, it lay yet
farther south.

But Ptolemy's vision of a fertile, populated southern land continued to
inspire, and in 1504, French explorer Captain Binot Paulmyer de Gonneville
claimed to have discovered it. He and his crew tarried six months in what de

Gonneville called "Southern India," refitted their ship, and enjoyed the hospitality of the land's inhabitants, who, he wrote, "asked nothing but to lead a life of contentment, without work." In July, de Gonneville departed this enchanted land for France, taking with him skins, feathered ornaments, and an "Indian" prince called Essomeric. The land was not, of course, the southern continent, and Essomeric was not an Antarctican. De Gonneville had probably been to southern Brazil. But it was an honest mistake, and his tales fed the public imagination and a growing enthusiasm for the existence of a southern Shangri-La.

Fifteen years after de Gonneville returned to France, Ferdinand Magellan became the first European to sail around the tip of the Americas, passing from the Atlantic to the Pacific through the strait that was subsequently named in his honor. To the south, he spied land—inhabited land at that, where people lived on cold, forbidding, mountainous terrain and kept fires burning day and night to protect themselves from the elements. In honor thereof, Magellan named this place Tierra del Fuego, or Land of Fire; but as he peered across the water and through the swirling mist, he could not tell whether it was a solitary island, a group of islands, or, perhaps, the tip of a great southern continent stretching farther south than the eye could see.

More than fifty years passed, during which time Magellan's discoveries remained unconfirmed by other European explorers; but the prospect of finding, claiming, and trading with such lands came to captivate the court of England's Queen Elizabeth I. Her geographers were strong believers in the lush, inhabited paradise of Ptolemy and de Gonneville, and Elizabeth was surely enticed by the thought of stealing it away from her European rivals—particularly given that a papal decree in 1493 had essentially declared that all new lands henceforth discovered be divided between Spain and Portugal. And so, in 1577, Elizabeth dispatched her favorite buccaneer, Francis Drake, in search of Magellan's discovery, with the instruction that any land "not in the possession of any Christian Prince" be claimed for England.

In August 1578, Drake's ship the *Golden Hind* arrived at the Strait of Magellan, slipped through it in just sixteen days (Magellan had taken thirty-seven), and sailed into the Pacific. But the ship ran afoul of the storms for which the region is justifiably renowned. Francis Fletcher, Drake's chaplain, recorded that

the winds were such as if the bowels of the earth had set all at
liberty; or as if the clouds under heaven had been called
together; to lay their force on that one place: the seas, which by
nature and of themselves are heavy, and of a weighty substance,
were rowled up from the depths, even from the roots of the
rocks . . . and being aloft were carried in most strange manner
and abundance . . . to water the exceeding tops of high and
lofty mountains.

The ship was swept farther south than any had been before, south of Tierra
del Fuego to the latitude of 57°S. Hereabouts, Drake and his crew fell upon a
group of islands, the identity of which has never been definitely ascertained but
which Drake named the Elizabeth Islands. Making landfall on one of them,
Drake lay down on his stomach, his arms outstretched over the cliff, and
declared himself the "southernmost man in the world." Had he been so
inspired, Drake might have been able to push farther south, perhaps becoming
the first man south of the Antarctic Circle. But despite the near-mythological
status his memory invokes in his homeland, and notwithstanding his instructions to find the southern continent for England, Drake was little more than a
pirate. More than anything, he sought Spanish gold, and so, resolving "not to
sale any further toward the pole Antarctick . . . and neerer the cold," he
turned north and east, sailing up the western coast of the Americas and across
the Pacific in search of galleons to plunder.

⁎ ⁎ ⁎

As Drake exemplified, the motivation of those who set out on expeditions and
those who dispatched them was rarely the pure pursuit of knowledge. Almost
invariably, it was a desire to conquer and colonize, to claim lands full of valuable resources that could earn riches for their discoverers and their discoverers' home countries, to convert the heathen masses to Christianity, and to
secure territories and thus control over trade routes and strategic straits. Any
inconvenience posed by the presence on such lands of indigenous populations
was sidestepped by the rationalization that such natives were not worthy of
being considered equals, a stance formalized in the fifteenth century by the

concept of *terra nullius,* whereby a population could be deemed too far from Christianity or civilization to be relevant.

It was just such a combination of considerations that in the 1760s provided the backdrop for the writings and proclamations of Alexander Dalrymple as he urged his countrymen to mobilize an expedition in search of the great unknown southern land—*terra australis incognita*—that he believed lay only just beyond England's grasp.

In the middle years of the eighteenth century, Britain was growing increasingly prosperous; although more than 75 percent of the population was still involved in working the land, a series of technological innovations was spurring the beginnings of the Industrial Revolution. The country was becoming a center for world trade, its overseas possessions providing a market for British textiles and metals and a source for commodities such as tea, coffee, sugar, and tobacco, which in turn were re-exported throughout Europe. By 1760, British overseas possessions and territories had taken 40 percent of English domestic products; in support of this growing world trade, the country's mercantile maritime fleet had mushroomed, from 250,000 tons in 1695 to 550,000 tons in 1760.

But Britain was not alone on the world stage. Although it won huge swaths of territory from France following the Seven Years' War of 1756–1763, its rivalry with its neighbor was far from over. There were conflicts and shifting alliances with other European countries as well; by century's end, Holland, Spain, and Portugal would all be at war with the English. Meanwhile, the country's subjects in the American colonies were chafing against the heavy-handed rule of a distant and decreasingly relevant monarch and would soon be in open revolt. Britain may have been powerful and its reach widespread, but its position was vulnerable, and there were those who were eager for it to find new lands and subjects to buttress its standing and wealth.

Among those was Alexander Dalrymple, for whom the situation was obvious: the security Britain sought could be guaranteed in perpetuity by finding and claiming the great southern continent. Who needed those 2 million or so ingrate colonials in America when in this great land there would be found "hospitable, civil and ingenious peoples," perhaps 50 million of them, with whom to do business? "There is at present no trade from Europe thither," he wrote, "though the scraps from this table would be sufficient to maintain the power, dominion and sovereignty of Britain, by employing all its manufactures and ships."

Dalrymple's journey to this promised land began with a naval career and a series of administrative postings, ranging from employ by the East India Company in Madras to the position of deputy governor of Manila. Possessed of a keen intellect and fueled by a desire to promote and expand British trade in the region, he immersed himself in the reports and firsthand accounts of Pacific exploration and discovery of new lands.

In 1767, shortly after returning to England following thirteen years abroad, Dalrymple collated his research into a volume titled *An Account of the Discoveries Made in the South Pacifick Ocean, Previous to 1764*. At the same time, he steadily began cultivating contacts and seeking to establish a position for himself among learned London society. What Britain needed, he urged, was another great explorer, another Columbus: someone with the knowledge, drive, and leadership needed to find the southern landmass and gather it safely into Britain's bosom. Someone, in short, very much like himself. All that was needed was the opportunity, the moment when the planets would align in the right pattern and present Dalrymple with the chance to meet his destiny.

The planets—or, specifically, one planet—presented him with just such an opportunity in 1769. One hundred thirty years earlier, a young astronomer named Jeremiah Horrocks had been the first to observe the transit of Venus across the face of the sun—a phenomenon in which the planet is directly between the sun and Earth, casting a black shadow on the star's surface and enabling astronomers and mathematicians, by observing from several points on the globe, to calculate the distance from Earth to Venus and also to the sun. The transit would next occur, it was calculated, on June 6, 1761, and then again on June 3 eight years later.

Although the 1761 transit was widely observed, the results were generally held to be scientifically disappointing. With the next transits not due until 1874 and 1882 and, thereafter, 2004 and 2012, scientists around the world resolved to mark the 1769 transit with the requisite degree of attention. At Britain's great seat of learning, the Royal Society, plans were set in motion to send teams of observers around the globe, including to the Pacific Ocean. The area of the

Pacific in which the observations were proposed to take place neatly over-lapped with the boundaries of Dalrymple's putative southern Shangri-La; and so, seeing his chance, Dalrymple leaped at it. He was, it so happened, well versed in astronomical matters. This, combined with his steady cultivation of the Royal Society and his self-publicizing as the country's authority on Pacific discoveries, placed him at the top of the Society's wish list for leaders of the voyage. The council of the Society formed a Transit Committee to make plans for the journey and noted that Dalrymple was "a proper person to send to the South Seas, having a particular Turn for Discoveries, and being an able Naviga-tor, and well skilled in Observation."

Dalrymple himself, with characteristic immodesty, acknowledged the honor the Society was bestowing upon itself by appointing him expedition leader. At the same time, he made it clear that his primary purpose was not so much to observe the transit of Venus as to set foot on the great southern con-tinent: "Wherever I am in June 1769 I shall most certainly not let slip an oppor-tunity of making an Observation so Important to Science as that of the Transit of Venus—I believe the Royal Society's Intentions make it unnecessary for me to say that there is but one part of the World, where I can engage to make the Observations."

Dalrymple's plan was set. The Royal Society had seen what to him was per-fectly obvious: that he and only he could take the helm of a voyage of such momentous importance. His place in the history books had been reserved. All that remained was to ink in the details of his entry.

But Dalrymple and the Royal Society had reckoned without the Navy. The Society had appealed to the Board of Admiralty for a ship, and even though the lords of the Admiralty had, before the Royal Society, seen the sense in making the Pacific voyage one that combined astronomy and discovery, they did not want Dalrymple, for all his protestations of unique qualifications, to be the per-son to lead it. The first lord of the Admiralty, Sir Edward Hawke, went so far as to state that he would see his right hand cut off before signing such a com-mission. The objection was not particularly a personal one; it was more a reac-tion to the idea of placing any civilian in charge of a naval vessel on a scientific voyage. It had happened before—when the noted astronomer Edmond Halley had led the *Paramore* on a voyage into the South Seas at the end of the seventeenth century—and, at least partly as a result of resentments toward Halley fostered by some fractious crew members, the experience had not been an altogether pleasant one. Hawke had vowed that it would not happen again.

Had it not been for his acerbity, Dalrymple might still have been able to play a major role in the expedition. The Society wanted him to take part in the voyage; he could have been the scientist on board and accrued all the plaudits that would have ensued on the return home should the journey prove as successful as he doubtless considered it would be with him involved. But as far as Dalrymple was concerned, it was all or nothing. It "may be necessary to observe," he had pointed out before the Admiralty handed down its decision, "that I can have no thought of undertaking the Voyage as a Passenger going out to make the Observations, or on any other footing than that of having the management of the Ship intended for the Service." He may have thought he was strengthening his hand; considering himself indispensable to the expedition, he appeared to believe he was in a position to make any demand he wished. Instead, he served only to box himself in. The Admiralty would not be moved. Just as his great ambition was within his grasp, it was denied him.

In his place, the Royal Society agreed to an approach from a twenty-five-year-old member, Joseph Banks, to travel on the voyage. The man selected to captain the ship and be in overall command of the expedition was a forty-year-old lieutenant from northern England who had recently distinguished himself by helping to wrest Quebec from the French during the Seven Years' War and subsequently charting the newly acquired territory of Newfoundland. He was also a skillful scientific observer, and this made his appointment as appealing to the Royal Society as it was to the Admiralty. His name was James Cook.

"A Country Doomed by Nature"

Cook's ship, the *Endeavour,* set sail from Plymouth on August 26, 1768. His instructions were to sail first to Madeira, to take in wine there, and thence to "proceed round Cape Horn to Port Royal Harbour in King Georges Island [Tahiti]," to observe the transit of Venus. "When this Service is perform'd you are to put to Sea without Loss of Time, and carry into execution the Additional Instructions contained in the inclosed Sealed Packet."

Just a few months previously, some of the crew of another ship, the *Dol-*

phin, believed that they had seen, to the south of Tahiti, mountaintops in the far distance. The Admiralty, anxious to explore the possibility that these mountains were part of the southern continent, factored the possible sighting into Cook's sealed instructions for the second part of the voyage, which dealt almost exclusively with this as yet undiscovered land:

> Whereas the making Discoverys of Countries hitherto unknown, and the Attaining of Knowledge of distant Parts which though formerly discover'd have yet been but imperfectly explored, will redound greatly to the Honour of this Nation as a Maritime Power, as well as to the Dignity of the Crown of Great Britain, and may tend greatly to the advancement of the Trade and Navigation thereof; and Whereas there is reason to imagine that a Continent or Land of great extent, may be found to the Southward of the Tract lately made by Captn Wallis in His Majesty's Ship the Dolphin (of which you will herewith receive a Copy) or of the Tract of any former Navigators in Pursuits of the like kind. . . . You are to proceed to the southward in order to make discovery of the Continent abovementioned until you arrive in the Latitude of 40°, unless you sooner fall in with it. But not having discover'd it or any evident signs of it in that Run, you are to proceed in search of it to the Westward between the Latitude before mentioned and the Latitude of 35° until you discover it, or fall in with the Eastern side of the Land discover'd by Tasman and now called New Zealand.

If Cook did discover the southern continent, the instructions continued, he was to carry out extensive exploration: mapping the coasts; documenting and, as appropriate, collecting examples of the area's flora, fauna, and mineral wealth; making contact with the local inhabitants, if any; and, "with the Consent of the Natives to take possession of Convenient Situations in the Country in the Name of the King of Great Britain; or, if you find the Country uninhabited take Possession for His Majesty by setting up Proper Marks and Inscriptions, as first discoverers and possessors."

Should, however, Cook be unable to locate the southern land, there was at least ample evidence, from the voyages of Dutch explorer Abel Tasman, that New Zealand existed—if not as part of *terra australis,* then as a land unto itself. Cook was to explore and map the coastline of this country, as much as the condition of his vessel, "the health of her Crew, and the State of your Provisions will admit of," before returning to England.

The *Endeavour* returned home in July 1771, its almost three-year voyage widely acclaimed a tremendous success. It was considered a success not because Cook had found the southern continent—he could hardly have been expected to, given that it was not to be found in the region he was directed to explore—or even because of the observations made of the transit of Venus, which were inconclusive. But the expedition's achievements were legion: Cook thoroughly charted the coasts of New Zealand, demonstrating conclusively that it was not a part of the southern continent. He sailed between New Zealand and the eastern coast of Australia and charted the strait between Australia and New Guinea. The *Endeavour* brought home 17,000 plants of 1,000 species, from New Zealand, Australia and elsewhere, that had never before been seen in Britain or Europe. Politically, too, it had been a tremendously successful voyage. Cook and his crew had spied, visited, or landed on nearly forty hitherto undiscovered islands and brought more territories into the British orbit at a time when, unlike its rivals such as the Netherlands and France, the country was starved for territories in the Southern Hemisphere.

Not everybody was impressed. Dalrymple, though mindful not to castigate Cook directly, nonetheless took the Admiralty to task for the fact that not only the voyage of the *Endeavour* but also several other recent British expeditions to the region had failed to find evidence of his giant southern continent. "I am very far from intending the most distant insinuation of resentment to, or dissatisfaction with, the worthy and brave old Officer who was at the head of the Admiralty when the Endeavour was purchased," he wrote. But, he continued,

> his open, honest, unsuspecting nature, I think, exposed him to
> the insinuations of *cunning* men, who would have endeavoured
> to throw the *odium on me* if the expedition, in the mode it was
> proposed, *had not been successful,* and attributed *all the merits* to
> their own tools. The point is not yet determined *whether there is*

or is not a SOUTHERN CONTINENT? Although four voyages have
been made *under their auspices,* at the same time I dare appeal,
even to them, that I would *not have come back in Ignorance.*

Given the size of the continent envisaged by Dalrymple, it would indeed
have required incompetence of the highest order to continually miss it. A more
self-critical individual might have considered the possibility that his predictions
were in error. But Dalrymple was likely to be impressed only by proof that the
continent existed, not by evidence of its absence. Yet the latter was equally
valuable, and in that sense Cook's voyage had been a triumph indeed. He had
sliced huge areas off Dalrymple's postulated southern landmass, establishing
fairly definitively that the long-held visions of a pan-hemispheric tropical paradise belonged in the realms of myth.

Dalrymple was not the only one who believed that questions remained
unanswered. "That a Southern Continent exists," wrote Joseph Banks, "I firmly
believe." He acknowledged, however, that it would have to be situated in much
higher latitudes—that is to say, closer to the South Pole—than had previously
been considered. Cook was more dubious, although he, too, conceded the possibility of it existing in very high latitudes. Nonetheless, he wrote, "I think it
would be a great pitty that this thing which at times has been the object of many
ages and Nations should not now be wholy clear'd up, which might very easily
be done in one Voyage without either much trouble or danger or fear of mis-
carrying as the Navigator would know where to go look for it." In other words,
Cook himself had gone a long way toward determining where the southern
continent was not; now he wanted to finish the job and establish whether it
existed at all.

The Admiralty concurred and, on September 25, 1771—with the *Endeavour* earmarked for other work—instructed the Navy Board to purchase two
vessels of about 400 tons for service in remote parts of the globe. The board
duly acquired two colliers, the *Marquis of Granby* and the *Marquis of Rockingham.*
These were renamed the *Drake* and the *Raleigh,* but amid concern that such
names might offend the Spanish, with whom relations had lately, if temporarily, been patched up, the names were changed again. The larger of the two, 111
feet long and weighing 462 tons, was named the *Resolution* and would be com-

manded by Cook; the smaller, at 97 feet and 340 tons, was christened the *Adventure* and placed under the command of Tobias Furneaux.

Overall command of the expedition would be Cook's. His plans were carefully laid out:

> Upon due consideration of the discoveries that have been made in the Southern Ocean, and the tracks of the Ships which have made these discoveries; it appears that no Southern lands of great extent can extend to the Northward of 40° of Latitude, except about the Meridian of 140° West, every other part of the Southern Ocean have at different times been explored to the northward of the above parallel. Therefore to make new discoveries the Navigator must Traverse or Circumnavigate the Globe in a higher parallel than has hitherto been done, and this will be best accomplished by an Easterly Course on account of the prevailing westerly winds in all high Latitudes. . . . [It] is humbly proposed that the Ships may not leave the Cape of Good Hope before the latter end of September or beginning of October, when having the whole summer before them may safely Steer to the Southward and make their way to New Zealand, between the parallels of 45° and 60° or in as high a Latitude as the weather and other circumstances will admit. If no land is discoveried in this rout the Ships will be obliged to touch at New Zealand to recruit their water.
>
> From New Zealand the same rout must be continued to Cape Horn, but before this can be accomplished they will be overtaken by Winter, and must seek Shelter in the more Hospitable Latitudes . . . after which they must steer to the Southward and continue their rout for Cape Horn in the Neighbourhood of which they may again recruit their water, and afterwards proceed for the Cape of Good Hope.

It was a bold and dramatic vision in which Cook, though expressing skepticism over the existence of the Grail for which he would search, nonetheless proposed doing nothing less than circumnavigating Earth, as much as was possible, around the latitude of 60°S. It would be a journey of truly epic proportions— one that surely, as Cook himself had suggested, would finally answer the question that had bedeviled nations and individuals for hundreds of years. As Cook expressed it, "If land is discovered the track will be altered according to the directing of the land, but the general rout must be pursued otherwise some part of the Southern Ocean will remain unexplored." If this voyage did not find the southern continent, the southern continent did not exist—or, if it did, it was only in still higher latitudes and certainly not nearly on the scale envisaged by Dalrymple.

* * *

Cook did not find Antarctica, although he came extremely close. On January 17, 1773, his ships became the first in history to cross the Antarctic Circle, but they were unable to push much farther south because of vast expanses of ice that blocked their way. Antarctica was just eighty miles distant, but Cook and crew could not know it.

Previous voyages had more than hinted at it, but Cook's expedition provided dramatic confirmation: not only was the southern continent, if it existed, far less extensive than Dalrymple and others had postulated, but it was also a long, long way from being a lush paradise inhabited by happy natives.

The small wooden ships were dwarfed by towering, terrifying icebergs, the like of which human eyes had rarely seen. Howling winds whipped icy spray onto the decks and turned the sails into giant sheets of ice. The Antarctic, it rapidly became clear, was a very unpleasant place to be.

The ships were bedeviled by fog. On February 8, an especially thick shroud caused the two vessels to become separated. After a further five weeks or so of sailing predominantly east at a latitude of about 60°S—all the while roughly paralleling the coast of Antarctica, some 250 to 300 miles away—Cook resolved to turn north and head for Australia and New Zealand.

By the next summer, he had returned. He crossed south of the Antarctic Circle on two more occasions—on December 20, 1773, and January 26, 1774. But on these voyages, too, he was denied a glimpse of the Promised Land. On

January 30, with the ship south of 70°S for the first time, the *Resolution* encountered another seemingly impenetrable wall of ice. Cook's journal tells the tale:

> The Clowds near the horizon were of a perfect Snow whiteness
> and were difficult to be distinguished from the Ice hills whose
> lofty summits reached the Clowds. The outer or Northern edge
> of this immence Ice field was composed of loose or broken ice so
> close packed together that nothing could enter it; about a Mile in
> began the firm ice, in one compact solid boddy and seemed to
> increase in height as you traced it to the South; In this field we
> counted Ninety Seven Ice Hills or Mountains, many of them
> vastly large. It was indeed my opinion as well as the opinion of
> most on board, that this Ice extended quite to the Pole or perhaps
> joins to some land, to which it had been wholy covered with Ice.

For Cook, this barrier was the final straw. His crew was tired and cold; the
Antarctic was an icy, foggy, hostile place to be; and by now he was certain that if
there was a southern continent anywhere to be found, it was not located in the
South Pacific—not unless it was much farther south. In fact, Cook and crew were
at this time barely ninety miles from the goal they sought; had they been at that
latitude or even farther north at a different longitude, they would have been in
no doubt that Antarctica did indeed exist. But for now, enough was enough:

> I will not say it was impossible anywhere to get in among this
> Ice, but I will assert that the bare attempting of it would be a
> very dangerous enterprise and what I believe no man in my sit-
> uation would have thought of. I whose ambition leads me not
> only farther than any other man has been before me, but as far
> as I think it possible for man to go, was not sorry at meeting
> with this interruption, as it in some measure relieved us from
> the dangers and hardships, inseparable with the Navigation of
> the Southern Polar regions.

Cook turned north. There was further exploration to be done of the South
Pacific, and Cook intended also to have one final sweep in search of the southern continent, this time in the South Atlantic, en route back to England.

On January 14, 1775, during this final sweep, the *Resolution* sighted a
mountainous, rocky, snow-covered crop of land in the middle of the ocean.
Through the snow and mist, it was difficult to ascertain its size and whether it
was merely one point on a larger landmass. On January 17, the ship pulled into
a bay and Cook took to a boat to investigate further:

> The head of the bay, as well as two places on each side, was ter-
> minated by a huge Mass of Snow and ice of vast extent, it
> shewed a perpendicular clift of considerable height, just like the
> side or face of an ice isle; pieces were continually breaking from
> them and floating out to sea. A great fall happened while we
> were in the Bay; it made a noise like Cannon. The inner parts of
> the Country was not less savage and horrible: the Wild rocks
> raised their lofty summits till they were lost in the Clouds and
> the Vallies laid buried in everlasting Snow. Not a tree or shrub
> was to be seen, not even big enough to make a tooth-pick.

Barren and bleak as this new land was, Cook and his crew at least had a
sense of optimism. Had they finally, after all this time, stumbled across the legendary southern continent? Alas, no. Cook sailed southeast, and on rounding a
promontory that he would later evocatively call Cape Disappointment, he
could see that the coast returned to the northwest, toward the point where he
and his crew had first set eyes on the land. It was an island, which he named
South Georgia. Nonetheless, he wrote, "The disappointment I now felt did not
affect me much, for to judge of the bulk by the sample it would not be worth
the discovery."

Indeed, by now, Cook's impression of the entire region was clearly one of
general disfavor. Commenting on the prospect of future Antarctic discovery, he
reflected:

> I firmly believe that there is a tract of land near the Pole, which
> is the source of most of the ice which is spread over this vast
> Southern Ocean. . . . It is however true that the greatest part
> of this Southern Continent (supposing there is one) must lay
> within the Polar Circle where the Sea is so pestered with ice,

that the land is thereby inaccessible. The risk one runs in exploring a coast in these unknown and Icy Seas, is so very great, that
I can be bold to say, that no man will ever venture farther than
I have done and that the lands which may lie to the South will
never be explored. Thick fogs, Snow storms, Intense Cold and
every other thing that can render Navigation dangerous one has
to encounter and these difficulties are greatly heightned by the
inexpressable horrid aspect of the Country, a Country doomed
by Nature never once to feel the warmth of the suns rays, but
to lie for ever buried under everlasting snow and ice.

But Cook underestimated his fellow men—and particularly their desire for
profit, even in the most unforgiving of conditions. Cook had definitively drawn
a black mark through Dalrymple's vision of a populated tropical Shangri-La,
but where he saw only "a Country doomed by Nature," others would see
opportunity. For, swimming in the seas and swarming over the bleak and "inexpressable horrid" beaches of South Georgia and the nearby South Shetland
Islands there were, noted Cook, huge numbers of seals. "The shores," he wrote,
"swarmed with young cubs," and it was all but impossible for a man to get
ashore among them.

The seals in question were fur seals, and their rich, thick pelts offered the
prospect of wealth for those who could collect them and sell them on the market—particularly in China, where they were much in demand. On beaches and
islands throughout the Northern Hemisphere and increasingly south of the
equator, the animals' fur—so vital to their survival in the wild—had spelled
their demise. Now, with news of Cook's sighting, the same would soon be true
of those in the Southern Ocean.

"The Ruthless Spirit of Barbarism"

It is not known exactly when the first sealing vessels reached South Georgia
Island, but the ink had barely dried on Cook's account of his voyage before the
island he claimed for Britain was crawling with sealers from England and
America. By 1800, seventeen American and British vessels were working the
area, that year taking a recorded 122,000 fur seals; one vessel alone, the *Aspa-*

sia, commanded by Edmund Fanning, sailed for home with the pelts of 57,000 seals in its hold. Sealer and explorer James Weddell estimated that by 1822, at least 1.2 million fur seals had been killed on the island.

The killing was brutal and efficient. When a seal beach was discovered, sealers would come ashore and lay waste with abandon. A seal would be struck hard on the nose with a club, turned over, and sliced open from head to tail, the knife pausing to stab its victim in the heart to kill it. The skins were stretched out to dry, salted, and, in the words of Amasa Delano—sealer and ancestor of Franklin Delano Roosevelt—stacked "in the manner of salted dried cod fish." Delano reckoned that an expert sealer could kill and skin sixty seals per hour. Each seal beach was worked relentlessly until, in the euphemistic words of one observer, it was "almost entirely abandoned by the animals."

When the fur seals had "almost entirely abandoned" South Georgia Island, the sealers turned their attention to the elephant seals that also lay along the beaches and that provided ample quantities of oil, ideal for lighting the lamps of city streets and houses back home. The killing of elephant seals was an ugly, bloody business. Sealers noticed that the enormous bulls, each of which commanded a harem of as many of sixty or seventy cows, kept a tight rein on the movements of each and every one of their females, refusing to let any out of their sight. This prevented the cows from sliding into the clutches of a rival, but it also proved an effective barrier to escape from the sealers, who would bludgeon the females first and then turn their attention to the bulls, prodding them until the animals reared up and opened their mouths in defiance and then shooting them in the palate and lancing them to make them bleed.

It was not long before fur seals and elephant seals alike were in serious decline on South Georgia Island. Weddell noted, for example, that "these animals [were] now nearly extinct." And so the sealers moved on, sailing ever farther south in their search for new beaches to ply their trade—in the process discovering new islands and filling in the empty spaces on the charts of the Southern Ocean. Exploitation spawned exploration and vice versa, with one almost accidental discovery in particular tipping off the sealers to the richest hunting grounds thus far and placing both hunters and explorers on Antarctica's doorstep.

Early on the morning of February 19, 1819, the cargo ship *Williams,* en route via Cape Horn to Valparaíso, Chile, had been blown off course by gale-force winds. Squinting through a series of snow squalls, William Smith, the ship's twenty-eight-year-old master, saw what he thought was mountainous, snow-covered land ahead; uncertain of what he was seeing, he hauled off and waited for the weather to clear. It did so the next day, and Smith's sighting was confirmed. With the aid of a chronometer, the use of which was gaining currency among mariners of the time, he calculated his location at approximately 67°S and 60°W and set course for Valparaíso.

On arrival in Chile, Smith reported what he had seen to William Shirreff, the senior Royal Navy officer in the region; Shirreff listened to his story and promptly suggested that what Smith had taken for land was merely an iceberg or two. Nonsense, protested Smith; he had sailed frequently in Greenland waters before now and surely could be trusted to tell the difference between dry land and a chunk of ice. Shirreff remained unmoved.

Perhaps this airy dismissal rankled and motivated him, but for whatever reason, after Smith left Valparaíso for his next destination, Montevideo, he decided while sailing back around Cape Horn to push south and try to find his land again. But winter was closing in, and the rapidly freezing sea soon forced him to abandon his efforts. In Montevideo, however, he found that reports of his discovery—and, most significantly, of the fact that he and his crew had seen prodigious numbers of seals and whales around the land he had spied—had preceded him. A brace of American sealers and merchants dangled money under his nose for details of his new land's whereabouts, but Smith refused their enticements, preferring—despite having been snubbed by his monarch's representative in Valparaíso—that priority and any accompanying riches fall to his homeland.

Three months later, sufficient cargo secured for another journey to Chile, Smith again set out for the treacherous journey around the Horn. Once more he steered off course, in search of his prior discovery. On October 15, the *Williams* came across an island, now known as Desolation Island, six miles from the previous sighting; the following day, they made landfall and staked a claim on behalf of King George III. If Smith had been contemplating the possibility that he had discovered the southern continent, he realized by the following day that he had instead encountered a series of islands—the South Shetland

Islands—and not one long coastline. It was a notable discovery nonetheless. The islands were two degrees farther south than the southernmost land yet discovered, and, as Smith noted, the waters and beaches contained "vast quantities of seals, whales."

This time, Smith's reports were taken more seriously in Valparaíso, and the Royal Navy determined to charter the *Williams*—with Smith sailing as master and pilot under overall command of a Navy man, Edward Bransfield—to ascertain whether he had indeed discovered a series of islands or whether his land was connected to a continent. On January 16, 1820, the *Williams* made its first landfall in the South Shetlands, sailing into Barclay Bay on Livingston Island; six days later, the ship reached King George Island, where the crew planted a Union Jack and asserted British sovereignty.

On January 30, forced to chart a course almost due south following a sudden adverse change in the weather, the *Williams* came across Tower Island, rising a thousand feet in the air; behind it, it seemed, there was yet more land. The ship sailed to the west of Tower Island and then south so the crew could better examine the vista that lay before them. It was mountainous, barren, bleak, icy, and cold. It presented, in the words of one crew member, "a prospect the most gloomy that can only be imagined." There was, he continued, only one thing that could redeem this forbidding horizon: "the idea that this might be the long-sought Southern Continent."

And that is exactly what it was. Bransfield, Smith, and the crew of the *Williams* were looking at the northernmost reaches of the Antarctic Peninsula. They had found Antarctica.

In an extraordinary twist, however, theirs was not quite the first sighting. For thousands of years, Antarctica had been the subject of curiosity and speculation, unseen by human eyes. Suddenly, the curtain had risen, and when it did so, the scene that had lain hidden for so long was revealed almost simultaneously to two separate audiences. For even as Bransfield and Smith gazed at the grim horizon of what they called Trinity Land, two Russian ships, the *Vostok* and the *Mirnyi,* were plying Antarctic waters on their own voyage of discovery.

The commander of these two ships, Fabian Gottlieb von Bellingshausen, was not motivated by the prospect of finding seals or whales, nor was he a merchant who had simply been blown off course. Bellingshausen was an ardent

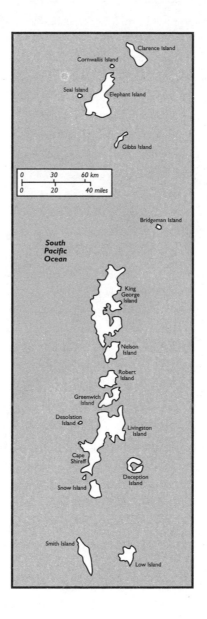

Clarence Island

Cornwallis Island

Seal Island

Elephant Island

Gibbs Island

| 0 | 30 | 60 km |
| 0 | 20 | 40 miles |

Bridgeman Island

South
Pacific
Ocean

King
George
Island

Nelson
Island

Robert
Island

Greenwich
Island

Desolation
Island

Livingston
Island

Cape
Shireff

Deception
Island

Snow Island

Smith Island

Low Island

admirer of James Cook, and his expedition, which had been authorized and encouraged by Emperor Alexander I, set out with the express purpose of exploring those regions Cook had missed. After three weeks in Rio de Janeiro, Bellingshausen sailed forth on his first southern season, in November 1819. He headed for South Georgia, complementing Cook's efforts by mapping the southern coast, whereas Cook had charted the north; continued on to examine the South Sandwich Islands; and then headed east along the sixtieth parallel before turning south and crossing the Antarctic Circle on approximately January 26, 1820.* At around midday on January 27, wrote Bellingshausen:

> . . . we encountered icebergs, which came in sight through the falling snow looking like white clouds. We had a moderate north-east wind with a heavy swell from the north-west and, in consequence of the snow, we could see for but a short distance. We hauled close to the wind, we continued on a south-east course and had made 2 miles in this direction when we observed that there was a solid stretch of ice running from east through south to west. Our course was leading us straight into this field, which was covered with ice hillocks.

Bellingshausen gave no impression of realizing exactly what he had seen, but today it is widely believed that he and his crew were looking at the ice shelves that rim what is now known as the Princess Martha Coast of Antarctica. An ice shelf may seem somewhat insubstantial in comparison with the mountains sighted by the crew of the *Williams,* but such shelves are an integral part of the Antarctic continent. Assuming we accord them the same status as rock and earth, Bellingshausen was the first to set eyes on Antarctica, three days before Smith and Bransfield.

---

*Although he did not himself highlight the crossing of the Circle as such, Bellingshausen did note his latitude and longitude on this date. It should be noted, however, that in accordance with the Julian calendar in use in Russia at the time, he recorded the date as January 15. Those dates are here converted in accordance with the modern Gregorian calendar to allow for better comparison with the achievements of Smith and Bransfield.

* * *

In a strange way, it is somewhat fitting that Bellingshausen should have apparently missed completely his discovery of the Antarctic mainland. As alluring as the prospect of finding *terra australis incognita* had been to thinkers and explorers over the centuries, the moment itself was anticlimactic. The notion of great riches and bounteous resources, which had led to so much licking of the lips on the part of Alexander Dalrymple and those before him, had been largely dashed by James Cook, and the crew of the *Williams'* lack of enthusiasm for the sight that befell them served only to underscore the continent's uninviting nature. Indeed, in the end, the return of the *Williams* and its sighting of Antarctica was overshadowed by the actions of other visitors to the area and the valuable cargo they secured.

Unknown to Smith, Bransfield, and crew, even as the *Williams* was sailing into Barclay Bay, sealers were already hard at work just five miles away. Smith's reluctance to part with information regarding the location of his find had proved little deterrent, and the determined and resourceful sealers had found their way there without his help. First on the scene, the British ship *Espirito Santo* spent thirty-three days on Rugged Island while its crew slaughtered vast numbers of seals. It was soon joined by the *Hersilia,* out of Mystic, Connecticut, a 131-ton brig on its maiden voyage. In just sixteen days, the *Hersilia* bagged almost 9,000 seals; it could have taken more, but it ran out of salt for preserving the skins.

In the end, it was Smith's discovery of the South Shetlands, not his subsequent sighting of Antarctica, that had the greatest impact. Wave after wave of sealers headed for the islands—indeed, shortly after returning to Valparaíso, Smith himself returned on a sealing voyage—until, within a few short years, the inevitable had happened. Visiting the islands in 1829—just ten years after Smith's discovery—W. H. B. Webster, surgeon on board HMS *Chanticleer,* noted:

> The harvest of the seas has been so effectively reaped that not a
> single fur seal was seen by us during our visit to the South Shet-
> land group; and although it is but a few years back since count-
> less multitudes covered the shores, the ruthless spirit of bar-
> barism slaughtered young and old alike, so as to destroy the race.

* * *

Dalrymple was wrong. There were no lush jungles, no friendly, hospitable natives. There were no resources for Britain to claim. There were ice, rock, and snow; fearsome storms; a raging ocean; and a bitter cold of the like few had ever endured.

There were fur seals—or, at least, there had been—and when their populations had been all but wiped out, interest in the region declined. There were other, short-lived efforts to revive the sealing industry in the region later in the 1800s and again in the following century. There were further voyages of discovery by sealers and explorers such as Weddell, John Biscoe, Peter Kemp, and John Balleny and by national expeditions such as those led by Jules-Sébastien-César Dumont d'Urville, James Clark Ross, and Charles Wilkes. But by the middle of the nineteenth century, the excitement that had engulfed the Antarctic had all but faded away.

It would not be long before it returned. As the century drew to a close, a new wave of explorers began planning another series of assaults on the southern continent. Their battles with the hostile conditions that awaited them would be legion: so much so that the period in which their voyages took place—from roughly 1901 to 1922—has become enshrined in history as the "heroic age" of Antarctic exploration, and renowned for the fact that not all the heroes escaped with their lives.

Their quest was largely for knowledge and national glory, but there followed in their wake searchers of an entirely different kind. Before the twentieth century was one-third complete, the waters of the Southern Ocean would be crisscrossed by vessels in pursuit of a new quarry, and the scale of the slaughter would rival that visited on the fur seals of South Georgia Island and the South Shetlands. The plunder of the Southern Ocean was not yet complete.

Chapter 4

So Remorseless a Havoc

The Ross Sea is a gateway into the heart of Antarctica, a giant wound slicing deep into the frozen continent. Three hundred seventy thousand square miles in area, its horizon is dominated on the western side by the dramatic vista of the Transantarctic Mountains, a 1,900-mile range that separates East and West Antarctica. The mountains look almost too perfect, like a freshly painted scenic backdrop, and the clarity of the Antarctic air seems to somehow foreshorten their distance: from 50, even 100 miles away, they appear almost close enough to touch, at least within striking range of a small boat, so that if you set out in the morning you could be there by lunchtime.

The sea is named after its discoverer, James Clark Ross, who entered here in January 1841 while on a four-year voyage of discovery and exploration and established a record for the farthest south anyone had yet sailed. It was a record that would not—indeed, could not—be broken, except by a matter of degree, for if you push your way to the very end of the Ross Sea, you will have traveled as far south as it is possible to go in a boat.

For Ross, the initial elation of breaching the last barrier of sea ice at the mouth of the sea and entering into open water and an unknown world—"may it continue to the Pole or a continent discovered," wrote one of the crew excitedly—was tempered by the sight of the western coast, which he called Victoria Land. Ross had sought the South Magnetic Pole, but the mountains and the erratic readings of his compass told him that although it was close, it was some way inland and beyond his reach. No matter; he recovered soon enough, claiming the area for his country and framing it as a "way of restoring to England the honour of the discovery of the southernmost land, which had been nobly won by the intrepid Bellingshausen, and for more than twenty years retained by Russia." Besides, how was it possible to remain disappointed when he and his men could only gaze in wonder at the extraordinary landscape—the "extreme solitude and omnipotent grandeur," in the words of Dr. John Robertson, the expedition's chief surgeon—with which they were now presented?

There was more to come. On January 28, after continuing south overnight toward land that had been in sight since noon the previous day, they saw in the near distance a tall mountain. From its summit, it at first appeared, a fine snowdrift was blowing. To the astonishment of all on board, however, on closer inspection the sight revealed itself to be an erupting volcano. The volcano, on an island that now bears Ross' name, was dubbed Erebus and a smaller companion named Terror, in honor of the expedition's ships. In a letter to Britain's Prince Albert, Ross described the sight thus:

> The volumes of smoke which were occasionally ejected and in sudden jets, attained in a few seconds the altitude of 14 to 1700 feet above the crater . . . the smoke then gradually dispersed by the action of the wind, leaving the crater quite clear and filled with intensely bright flame, most distinctly visible notwithstanding the presence of the meridian sun. The perennial snow which clothes the whole of this land extended even to the verge of the crater, so that the flame and smoke appeared to issue from a monstrous iceberg, and if any lava was discharged it must have passed away beneath the snow, as it could not be distinguished by our best glasses.

The juxtaposition of this fiery spectacle and the frozen netherworld that otherwise surrounded them left many on board the *Erebus* and *Terror* humbled. Wrote scientist and assistant surgeon Joseph Hooker:

This was a sight so surpassing everything that can be imagined and so heightened by the consciousness that we have penetrated, under the guidance of our commander, into regions far beyond what was ever dreamed practicable, that it really caused a feeling of awe to steal over us, at the consideration of our comparative insignificance and helplessness, and at the same time an indescribable feeling of the greatness of the Creator in the works of his hand.

One can only guess at the thoughts that must have been pulsing through the minds of those on board both ships. The ice that seventy years before had been so thick as to deter and repel as great an explorer as James Cook was now behind them. They had entered a new realm—a world of fire *and* ice; what else might lie ahead?

It soon became clear that their voyage to the south would not last much longer. Recorded Ross:

> As we approached the land under all studding-sails, we perceived a low white line extending from its eastern extreme point as far as the eye could discern to the eastward. It presented an extraordinary appearance, gradually increasing in height as we got nearer to it, and proving at length to be a perpendicular cliff of ice, between one hundred and fifty and two hundred feet above the level of the sea, perfectly flat and level at the top, and without any fissures or promontories on even its seaward face. What was beyond it we could not imagine; for being much higher than our mast-head we could not see anything but the summit of a lofty range of mountains extending to the southward as far as the seventy-ninth degree of latitude. . . . It was . . . an obstruction of such a character as to leave no doubt upon my mind as to our future proceedings, for we might with equal chance of success try to sail through the Cliffs of Dover, as penetrate such a mass. When within three or four miles of this remarkable object, we altered our course to

the eastward, for the purpose of determining its extent, and not without hope that it might still lead us much farther to the southward. The whole coast here from the western extreme point, now presented a similar vertical cliff of ice, about two or three hundred feet high.

This wall of ice he named the Victoria Barrier, in honor of Britain's queen. Later, it would be called the Great Ice Barrier, the Great Southern Barrier, the Icy Barrier, the Ice Barrier, and simply the Barrier before being dubbed the Ross Ice Barrier and the Ross Barrier in honor of its discoverer. Now known as the Ross Ice Shelf, it is a phenomenon in itself. Covering roughly 200,000 square miles—an area a little less than that of Texas—it is the largest body of ice in the world. Its seaward edge stretches 400 miles from coast to coast, and the shelf extends 600 miles inland. The cliffs that barred Ross' passage were about 200 feet high, but the average thickness of the ice, which slopes up gently toward the South Pole, is roughly 1,100 to 1,200 feet. Anchored firmly to the continent, it is nonetheless unsupported by any rock or land and is therefore floating on the sea beneath it.

To the east of Ross Island, at slightly varying locations depending on the movement of the ice, there is an indentation in the Ross Ice Shelf. The famed explorer Ernest Shackleton dubbed this area the Bay of Whales, and in 1908 he considered using it as the base for his planned assault on the Pole before deciding that would be too dangerous. Shackleton ultimately settled on a northwestern corner of Ross Island called Cape Royds and from there launched an outstanding effort in which he and three companions reached the farthest point, north or south, yet attained by anybody. They closed to within just ninety-seven nautical miles of the Pole, but weakened by the profoundly difficult conditions, Shackleton recognized that even though they might still reach the Pole, the attempt would ensure that they would not return alive. Having decided, as he later explained to his wife, that he would rather be a live donkey than a dead hero, he turned around, barely making it back to Ross Island.

Three years later, however, Norwegian Roald Amundsen, perhaps the most accomplished polar explorer of his or any era, did elect to begin his polar attempt from the Bay of Whales. He had decided that the risks of placing camp on a shifting, relatively unstable surface were offset by the advantages proffered by the sixty-mile head start it would give him on a rival British expedition led by Robert Falcon Scott and based at McMurdo Sound. A clinic in efficiency, Amundsen's expedition set out from its base on October 20, 1911; by Friday, December 14, it had achieved its goal. Amundsen and his companions had become the first people to reach the South Pole.

A little more than a month later, Scott and four companions, believing they would be first, were devastated by the sight of a flag and a small camp and the knowledge that they had been forestalled by the Norwegians. Crushed by defeat, they set out on a desperate struggle back to their base camp. This race, too, they would lose: two of the five died en route; the bodies of the remaining three, including Scott, were found by a search party the following November. The search party retrieved the three men's personal effects, collapsed the tent in which they were found, and erected a cairn on top of it. Over the years, it was steadily covered with ice and snow and carried slowly forward by the inexorable northward march of the Ross Ice Shelf until one day, unseen by human eyes, it was presumably quietly surrendered to the Ross Sea.

Scott's assault on the South Pole was his second voyage to the Antarctic Regions. In early 1902, he entered the Ross Sea on board the *Discovery,* only the second expedition after Ross' known to have visited McMurdo Sound, and the first actually to set foot on Ross Island. Indeed, after Ross turned north and left his eponymous sea, no other vessel entered its waters for fifty-four years. That next arrival paid the briefest of visits: the ship—named, appropriately enough, the *Antarctic*—crossed south of the Antarctic Circle on December 25, 1894, sighted Cape Adare—the promontory that marks the Ross Sea's unofficial northwestern boundary—on January 16, 1895, and by February 8 had turned around and left Antarctica behind. But this appearance, although fleeting, was significant. For one thing, the ship was a steam vessel, one of the earliest to visit the Antarctic and a sign of things to come. And even though Scott, Amundsen, Shackleton, and others would bring the greatest amount of worldwide public attention to Antarctica for their efforts in battling the elements and attempting to peel away, layer by layer, the secrets of the frozen continent, this particular interloper was engaged on a mission the consequences of which would leave a significant and lasting impression on the environment of the Antarctic and

Southern Ocean. In the same way that, for all of James Cook's magnificent contributions to uncovering the mysteries of *terra australis incognita,* the most
immediate result of his discoveries was the massacre of the region's fur seals,
so were some Europeans who read the accounts of James Ross' voyage
inspired, above all else, to head south in search of whales.

* * *

By the mid- to late nineteenth century, commercial whaling in the Northern
Hemisphere was in decline. The massacre of right and bowhead whales had
been accompanied by the decimation of sperm and gray whales in the Pacific
Ocean and elimination of grays in the Atlantic. The increasing availability of
petroleum oil for lighting and heating cut into profits. Money was still to be
had, but the industry was becoming a lot less lucrative, and the incentives to
endure the enormous sacrifices involved in being at sea on a whaler for months
or years on end were becoming far smaller. Between 1850 and 1872, the number of ships involved in the global whaling industry had dropped from 400 to
72.

   One thing above all others kept the industry going through the lean years:
the overwhelming urge of women to squeeze their bodies into tiny corsets
made from baleen—and the even greater desire of men to look at women's
bodies squeezed into tiny corsets made from baleen. And even if the whale
stocks of the Northern Hemisphere had been decimated, Ross' accounts of his
voyages fostered images of Southern Ocean waters practically boiling with
thousands upon thousands of whales, just waiting to roll over and die for England.

   On January 14, 1841, two weeks before coming across the erupting Mount
Erebus, Ross noted:

> A great number of whales were observed; thirty were counted
> at one time in various directions and during the whole day,
> wherever you turned your eyes, their blasts were to be seen.
> They were chiefly of large size, and the hunch-back kind: only
> a few sperm whales were distinguished amongst them, by their
> peculiar manner of "blowing" or "spouting" as some of our men

who had been engaged in their capture called it. Hitherto, beyond the reach of their persecutors, they have here enjoyed a life of tranquility and security; but will now, no doubt, be made to contribute to the wealth of our country, in exact proportion to the energy and perseverance of our merchants; and these, we well know, are by no means inconsiderable. A fresh source of national and individual wealth is thus opened to commercial enterprise, and if pursued with boldness and perseverance, it cannot fail to be abundantly productive.

Optimism unbound, and repeated in observations he made on the other side of the continent the following year while sailing off Joinville Island, at the northeastern end of the Antarctic Peninsula:

We observed a very great number of the largest-sized black whales, so tame that they allowed the ship sometimes almost to touch them before they would get out of the way; so that any number of ships might procure a cargo of oil in a short time. Thus within ten days after leaving the Falkland Islands, we had discovered not only new land, but a valuable whale-fishery well worth the attention of our enterprising merchants, less than six hundred miles from one of our own possessions.

Despite Ross' patriotic urgings, his countrymen were not swift to respond to the challenge. However, in 1873, a German expedition headed south on board the *Grönland* in search of the whales Ross had spied. The *Grönland* made history as soon as it crossed the Antarctic Circle: it was the first steamship to do so, and as had many sealing expeditions before it, its voyage added new pieces to the puzzle of Antarctic cartography—in this case, the Bismarck Strait and the Neumayer Channel in the Antarctic Peninsula region. There were some financial benefits, too, in the form of sealskins and oil. But in terms of whales, the expedition was a disaster. The reason, simply put, was that its backers assumed that the "largest-sized black whales" Ross had recorded in the Weddell Sea region were right whales. There were whales, sure enough, and plenty of them—however, they were not the slow, methodical right whales but what scientists refer to as rorquals—blue whales, fin whales, sei whales, Bryde's whales, humpback whales, and minke whales.*

Rorquals have a number of physiological features in common that they do not share with, for example, right and bowhead whales. They are generally sleeker and less bulky. Their principal distinguishing characteristic is pleated throat grooves that allow them to greatly distend their throats and consume vastly more food in a single gulp than would otherwise be the case. There was nothing inherently undesirable about rorquals to the whalers of the nineteenth century, although their blubber was thinner and their baleen plates shorter; the problem was that they were too strong, were too fast, and, unlike rights and bowheads, did not float when killed. Lowering small boats into icy seas to fire harpoons into the backs of rorquals was simply not a viable proposition.

But lingering good stories and desperation have fueled many a risky venture, and in 1874 the Gray brothers of Peterhead, Scotland, issued a pamphlet titled *Report on New Whaling Grounds in the South Seas* in an effort to drum up sufficient interest to finance a startup of Antarctic whaling. Given the brothers' enthusiasm for the idea, the pamphlet was predictably bullish about its prospects, citing the observations of wall-to-wall whales Ross and others had made and determining, notwithstanding the absence of any evidence one way or the other, that the whales Ross and company had seen were indeed right whales, not rorquals. The pamphlet elicited much interest; nonetheless, it was some time before the necessary sponsors were found. Finally, one Robert Kinnes of Dundee set about equipping a fleet of four ships with a total crew of 130. In September 1892, the Dundee whaling fleet—the *Balaena,* the *Diana,* the *Active,* and the *Polar Star*—set sail for the Antarctic, reaching the western Weddell Sea on December 24.

One of those on board the *Balaena,* painter William Burn Murdoch, reflected on the venture with some cynicism:

> The *Balaena mysticetus,* right whale, Bowhead or Greenland whale, or whatever the reader may choose to call it, is, as he

---

*All these species are found in Antarctic waters except the Bryde's whale, which is restricted to the tropics and subtropics.

perhaps already knows, of great value on account of the bone in its mouth. You will find in the *Ladies' Pictorial* plenty of pictures of the people who keep up the price, or you can see them half alive in the streets—willowy things with their blood all squeezed into their heads. The whalebone in the jaws of one whale sometimes is worth two or three thousand pounds. Naturally, a whale with such a fortune in its mouth has been in great request, and in consequence has become so scarce, or so retiring, that of late years Arctic whalers have found their formerly profitable industry almost a failure.

To make a new start, the Nimrods of Peterhead, three brothers Gray, of Arctic fame, proposed taking their ships to the Antarctic to look for whales there. From the account given by Sir James Ross of his voyage of discovery in 1842, there was reason to believe that all that was necessary to make a 'full ship' was to sail south, haul the bone aboard, and sail home again with a fortune between decks. Glorious castles in the air were built in this prospective foundation of bone and blubber.

Castles in the air, indeed: constructs so insubstantial and inaccurate that in the words of the scientist on board the *Balaena,* one William Bruce—who would later achieve fame as leader of a highly successful Scottish Antarctic expedition—none of the vessels "saw any sign of a whale in the least resembling the Greenland or Bowhead whale."

So desperate did the members of the expedition become, it seems, to harpoon *something,* that at one point a boat from the *Active* made fast to one of the seemingly innumerable fin whales that lay lazily at the surface around them. Murdoch recorded the result:

> The whale went off in a bee-line with the three lines in the boat; a second boat followed and made fast, and again the whale made off, with three more lines, that is, with 720 fathoms in all, or 1440 yards of two-inch rope trailing behind it. . . . To save the first lines a third boat fired another harpoon into the whale, and this time the line was brought on board ship; and the ends of the

other two lines were also picked up and brought on board, and away went the procession in tow of the whale, the three boats hanging astern! . . . The whale towed them along at a good rate, and rockets were fired into it whenever there was a chance, but it only showed its nether end above water, so their effect was only to make it go faster. After fourteen hours' play the engines were reversed and the lines broke, and the whale went away 'with half Jock Tod's smithy shop in its tail.'

And that was as close as any of the Dundee whaling fleet came to catching a whale. Nor were they alone: one of those sought out to finance the expedition, a Norwegian by the name of Christen Christensen, had responded to the request by equipping a ship of his own, the *Jason,* and appointing thirty-two-year-old Carl Anton Larsen as captain. The *Jason* met up with the Dundee ships on several occasions, the crews rowing back and forth for discussions both professional and social. Eventually, the captains of the *Balaena, Diana, Polar Star,* and *Jason* met together and agreed: it was time to cut their losses, kill as many seals as they possibly could to earn at least some money, stop off at Montevideo or some other port to pick up salt to preserve the sealskins, and then head home. A few months later, both expeditions returned to their home ports, their voyages financial disasters.

For the British, that, for a while at least, was that. Indeed, in 1893, several British scientists declared that there simply was no point in contemplating whaling in the Antarctic. There seemed to be only rorquals in the area, and rorquals were not only far harder to catch but also just not worth the effort: their baleen was too short, their blubber too thin. The Norwegians, on the other hand, were not so easily deterred, even though Larsen's return voyage in 1893–1894 was also without success. For eighty-four-year-old Svend Foyn, the gray eminence of Norwegian whalers and a pioneer who would ultimately transform, prolong, and expand the commercial whaling industry worldwide, the problem was simple: so far, everyone had been looking in the wrong place. All the expeditions had gone to the Weddell Sea region, yet some of Ross' most dramatic descriptions of whales had been from the sea that bears his name.

With financing by Foyn, an old steam whaler, the *Kap Nor,* was refitted and renamed the *Antarctic.* It carried eight whaleboats, thirteen harpoon guns, and a crew of thirty-one. The captain was Leonard Kristensen; the "manager" of the expedition, however, was an expatriate Norwegian and associate of Foyn's, Henryk Bull. Although Bull was ostensibly in overall command, his duties were poorly defined, and the expedition was plagued by personality conflicts between the two men. It was just one of the problems that were to afflict the *Antarctic.*

To begin with, the ship did not leave Norway until September 20, 1893, perilously late in the season. By the time it reached the Antarctic, the vessel had sprung a leak and begun to run low on coal; its attempts to break through the thickening pack ice outside the Ross Sea were in vain, and the ship retreated to the subantarctic Kerguelen Island—where it was nearly wrecked by wind squalls—before limping on to Melbourne, Australia.

Bull and Kristensen tried again. On September 28, 1894, the ship left Melbourne, destined for the Ross Sea. The ship's propeller came loose from its shaft, however, and the *Antarctic* had to hobble its way to New Zealand for repairs. Two crew members deserted, and another seven refused to go south and were discharged. During the return south, the remaining crew members attempted to shoot some blue whales but achieved little more than losing their harpoons. Finally, after breaking through the pack ice and entering the open water of the Ross Sea, the ship was able to linger for only a couple of weeks before once again running low on coal and heading back north.

The sole whaling success of the entire expedition took place during the winter following the initial return from Kerguelen, when the *Antarctic* finally came across some right whales off Campbell Island, 350 miles south of New Zealand. But the catching equipment failed completely, and the crew took only a solitary calf—the only right whale killed by any of the would-be whaling expeditions between 1892 and 1895.

Despite being as dismal a failure as its predecessors in direct commercial terms, the *Antarctic* did earn a place in history and open the doors for future whaling success. On January 24, 1895, a small boat lowered from the ship approached Cape Adare and one of its occupants stepped ashore, making the

first confirmed landing on the Antarctic mainland.* Just which occupant took
that first step remains a matter of some dispute. Captain Kristensen claimed it
was he, recording in his journal, "I was sitting foremost in the boat, and jumped
ashore as the boat struck, saying, 'I have then the honor of being the first man
who has ever set foot on South Victoria Land.'"That account was contradicted
by a young Norwegian, Carsten Borchgrevink:

> I do not know whether it was the desire to catch the jelly-fish
> (seen in the shallows), or . . . a strong desire to be the first
> man to put foot on this *terra incognita,* but as soon as the order
> was given to stop pulling the oars, I jumped over the side of the
> boat. I thus killed two birds with one stone, being the first man
> on shore, and relieving the boat of my weight, thus enabling her
> to approach land near enough to allow the captain to jump
> ashore dry-shod.

Whether it was Kristensen or Borchgrevink—or Alexander Tunzleman, a
crew member from New Zealand who also claimed priority—it was Bull who
noted that the expedition's most significant consequence was that it had

> proved that landing on Antarctica proper is not so difficult as it
> was hitherto considered, and that a wintering-party have every
> chance of spending a safe and pleasant twelvemonth at Cape
> Adare, with a fair chance of penetrating to, or nearly to, the
> magnetic pole.

Sure enough, in 1899 Borchgrevink returned to the area as leader of the
first expedition to winter over on the Antarctic mainland—although the South
Magnetic Pole would not be reached until 1909. Bull also drew another prescient conclusion from the wreckage of the *Antarctic* expedition. Noting that
there was no suitable island on which to establish a whaling station in the Ross
Sea area, he determined that any future whaling operations in the region would

—————

*There is evidence, however, that at least five earlier landings were made by sealers in the
Antarctic Peninsula region.

have to "consist of at least two vessels—the one a small steamer, to do the actual hunting, and the other, a store ship of fair tonnage."

That was exactly the form that Antarctic whaling would ultimately take. The man who would take it in that direction was none other than Carl Anton Larsen, who had been captain of the *Jason.* The ship that indirectly would lead him to discover the secret of whaling's success in the Southern Ocean was the *Antarctic.* And the adventure that culminated in this discovery was an expedition like no other, one of the greatest tales of adventure and survival ever told.

The Survival of Otto Nordenskjöld

With the decline of sealing in the Antarctic and subantarctic, exploration in the region had also taken something of a nosedive. For more than fifty years after Ross, the only nonwhaling voyage to the area was the scientific expedition of the *Challenger,* a worldwide voyage of oceanographic discovery that crossed the Antarctic Circle on February 16, 1874.

The Sixth International Geographical Congress met in London in July 1895 and passed a resolution that "further exploration of the Antarctic regions should be undertaken by the end of the century." The immediate result of this call was that the dawn of the twentieth century saw three roughly simultaneous expeditions heading toward the southern continent: a German voyage on board the *Gauss,* Scott's *Discovery* expedition, and a Swedish venture led by Otto Nordenskjöld. The ship used by the Swedish expedition was the same *Antarctic* that had searched for whales in the Ross Sea, and the captain was Carl Anton Larsen. The choice of Larsen was a good one, not only because he now had extensive experience in the Antarctic Peninsula region, to which the *Antarctic* was headed, but also because he had proven himself an accomplished explorer. During the first *Jason* voyage, he had landed on the eastern coast of Graham Land, on the Antarctic Peninsula, and uncovered some of the first fossils ever to be found in the Antarctic—petrified wood remains, evidence that Antarctica had once been a warmer, much more hospitable place.

The plan for the expedition was to find a suitable site for Nordenskjöld and a team to be set down to spend the winter. Meanwhile, the *Antarctic* would head to the Falkland Islands and pick up a young scientist called Gunnar Andersson, who would take over Nordenskjöld's duties as expedition leader. The ship would spend the rest of the summer and fall conducting scientific surveys and then return the following summer, pick up the wintering party, and head home.

On February 9, 1902, Nordenskjöld selected Snow Hill Island as the site for the winter camp, and after a few days of transferring supplies from ship to shore, he and five companions rowed to land as the *Antarctic* steamed off into the distance. Despite being pounded mercilessly by raging storms, the shore party survived the trials of winter without great mishap, and with the return of summer, they waited expectantly for the return of the *Antarctic*. As the days wore on and there was no sign of the vessel, the party grew nervous. Trips to the top of a nearby hill to search for a sign of the ship's approach filled them with increasing foreboding: day after day, their gaze was met only by vast expanses of ice-covered sea. It seemed difficult to imagine how the *Antarctic* would be able to force its way through that barrier, and sure enough, as weeks passed with no rescue in sight, the party conceded the inevitable and settled down for another winter, left only to dream sadly of family and friends back home and to pray that their ship might still return.

After leaving Nordenskjöld and his companions, the *Antarctic* had continued conducting research as planned in the region of the Falklands and South Georgia, and early on the morning of November 5, 1902, it set sail from Ushuaia, on the tip of Argentina, to relieve the shore party. It was not long, however, before the *Antarctic* encountered surprisingly heavy ice conditions. Noted Gunnar Andersson:

> Two days later, on the night between the 11th and the 12th, our further progress was stayed by the edge of close pack-ice, and, after making a few attempts at forcing it, we soon found ourselves fast in impenetrable ice. . . . As soon as the pack grew more open, Larsen rammed his way forward a bit, but the ice closed round us again. . . . During the course of these clear days we could stand on the bridge and count fifty icebergs of varying form around us.

Undeterred, they pushed on, continuing with their research and expecting to be at Snow Hill Island by about December 8. But as they approached the area from the northeast, they discovered what Nordenskjöld and his companions already knew: that Erebus and Terror Gulf, the stretch of water through which they had to sail to reach their expedition-mates, was nearly everywhere coated with solid pack ice. After several unsuccessful attempts by Larsen to force the *Antarctic* through, Andersson determined that the best option was for a small party to put ashore where they could and make an overland passage to Snow Hill Island. If the ship had not also reached Snow Hill Island by an appointed day, they would return with the overwinterers to their starting point, where, it was hoped, the *Antarctic* would be waiting.

But barely had Andersson and comrades set out from what they called Hope Bay when they found their way blocked by a channel of ice and, ironically, open water. Despite these obstacles, the extensive leads through the ice gave them hope that even if their path was closed off, the *Antarctic* would at least be able to find a way through after all.

Those hopes, too, were to be dashed. The team returned to their Hope Bay camp and settled down to wait for the ship to return. But as Andersson put it: "The days came and the days went, and weeks became months, but no *Antarctic* arrived. The necessity of wintering in this place—a thing we had at first discussed as a distant possibility— grew gradually to a threatening certainty." They set about building a winter hut; for several months they hunkered down in their new home, subsisting on penguin and seal meat and what little food they had brought with them and using blubber as fuel for their stove. It kept them warm, but it threw off a thick black soot that coated the walls of the hut and the faces, hair, and clothes of the people in it.

What they could not know was that shortly after putting them ashore at Hope Bay, the *Antarctic* had perished—crushed between ice floes—and its crew had been forced to scramble across twenty-five miles of shifting sea ice to the safety of Paulet Island. Antarctica had conspired against Nordenskjöld's expedition in every conceivable way. His plan now lay in ruins, and the expedition's members were scattered in three isolated groups, each unaware of the fate of the others, and all forced to survive through an Antarctic winter for which they were not prepared.

But if fate had intervened to wreck Nordenskjöld's plans, it worked in improbable ways to bring the expedition to an extraordinary conclusion. On

October 12, 1903, Nordenskjöld and Ole Jonassen, in the course of a sledge journey from Snow Hill, made a startling and unexpected discovery:

> We soon reach the cape mentioned, and I imagine for a moment that I can catch a glimpse of something of an unusual appearance, but pay no further attention to it, when Jonassen speaks again: "What's that strange thing there close by the land?" I glance thither and say: "Yes, it looks like men, but it can't be, of course; I suppose it is some penguins!" and continue to march onwards. But Jonassen says at once: "Hadn't we better stay so that you can see what it is?" For the third time I look at the objects in question: of a certainty they do look strange, and a feeling tells me that something of importance is there. I take my field-glass. My hand trembles a bit when I put it to my eyes, and it trembles still more when the first look convinces me that it is really men that I see. . . . I soon hear a faint cry, which I take to be an "hurrah!" I do not answer, for the matter is as yet all too mystical for me, and I can now see so much that I mark the strangeness of the figures that are coming towards us. It cannot be that these two creatures are of the same race of men who were once my companions on board the *Antarctic*. . . .
>
> And what is it I see before me? Two men, black as soot from top to toe; men with black clothes, black faces and high black caps. . . . Never before have I seen such a mixture of civilization and the extremest degree of barbarousness; my powers of guessing fail me when I endeavour to imagine to what race of men these creatures belong.

It was a while before Nordenskjöld fully grasped what was going on. The strange forms they had encountered were Andersson and one of the other men from the Hope Bay party, who had themselves struck out in another attempt to reach Snow Hill. This time successful in navigating the obstacles that had previously barred their way, they were still some distance from their destination, and it was only by the most extraordinary stroke of chance that they had encountered their lost comrades just where and when they did. It is a measure of how little was known about the Antarctic at that time that, before reality dawned, Jonassen had ventured that maybe the apparitions were some form of natives. But after uncertainty had given way progressively to shock, disbelief, and delight, the reunited parties returned to the base camp and waited for the rescue they hoped would come.

On the morning of November 8, the men saw more figures approaching along the ice in the distance. Excitedly, some rushed off to meet them as others waited anxiously at the camp. The automatic assumption was that the *Antarctic* had finally succeeded in making its way to Snow Hill, that they were all finally about to be saved, that the figures making their way across the snow in the distance were Larsen and some of the rest of the crew. Rescue, it turned out, was at hand, but not in the form of Larsen or the *Antarctic*. The approaching forms were those of Captain Julian Irizar and Lieutenant Jalour of the relief vessel *Uruguay,* which had been dispatched by the government of Argentina in the absence of news from the *Antarctic* or the rest of the expedition. At long last, they were saved; but, as Nordenskjöld wrote, this enormous relief was tempered by the revelation from Irizar and Jalour that no one had heard anything of the *Antarctic:*

> For the second time within a month we stand face to face with one of those moments when one's whole world of sense seems to revolve itself into a mist, in the presence of the all-subverting unexpected *new* which draws us into its vortex. We had been so convinced that at this time of the year no vessel but the *Antarctic could* come, that when we saw the party approaching, we did not for a moment doubt that it was our old companions coming towards us. Had we learned of the loss of the *Antarctic* in an ordinary way, the blow would have been a crushing one, and nothing more. But now the news, together with the knowledge that we were to be released in this unexpected way, and the thought of the enormous responsibility resting in the decision that would have to be made within the next few hours, seemed, for the moment, to deprive me of all power of motion.

Sorrow was depicted on every countenance, for everyone saw
how small was the hope of ever again seeing the comrades we
had left on board the *Antarctic*.

The day's dramas were far from done, however:

> None of us intended going to bed that night. Each one was
> working silently at his own tasks, and I sat myself at my desk
> . . . when we heard the dogs begin to bark and howl. . . . One
> of us went and cast a look through the opened door. When he
> returned, he said that there were people down on the ice—"Six
> or eight men, I think". . . .
>
> Out on the hill there was a group of men looking up at the
> flag which still waved above our house. . . . Is it an optical
> delusion, produced by the anxieties of the day, or is reality once
> more about to surpass all that expectation and imagination
> combined could ever picture? . . .
>
> The next moment, wild, ear-piercing cheers, mingled with
> shouts of "*Larsen!* LARSEN is here!!" tear us away in an instant
> from the work we have in hand. As a matter of fact, we have
> experienced so much during the last few days that nothing can
> seem impossible to us; but still, I can scarcely believe my ears.
> There must be some mistake; it must be the day's unrest that has
> made one of us give a form of reality to his wishes. But I hurry
> out like the rest, and the next instant all doubts are vanished.

In an extraordinary feat of timing, which would scarcely seem believable if
inserted into a movie screenplay, Larsen and five companions—who had set
out in an open boat from Paulet Island for Andersson's camp at Hope Bay, discovered the party's note that they were heading toward Snow Hill and then
themselves rowed and walked to the base camp—had arrived on the very same
day as the *Uruguay*. Had they been just a couple of days later, the *Uruguay* and
Nordenskjöld's party would have since departed, Larsen and the crew of the
Antarctic would surely have been doomed, and history might have turned out
differently. As it was, the *Uruguay* was able to pick up all the survivors from

Paulet Island, and the entire expedition was treated to a hero's welcome in Buenos Aires.

Larsen's response to this adulation appears to have been one of bemusement. Even after everything the various parties had endured over the previous eighteen months or more, despite the fact that only phenomenal luck had prevented Larsen and his crew from being trapped on Paulet Island and dying there, the phlegmatic skipper seemed not to have allowed himself to be caught up in the moment. Incredibly, even then, his thoughts turned to whales. Twice before, he had visited the region in search of them, without great success; this time around, while visiting the Falkland Islands and South Georgia Island after depositing Nordenskjöld and his party on Snow Hill Island, he had come across the mother lode. Perhaps it was the thought of all those whales that had sustained him during his enforced wintering; certainly it was the subject to which he steered everyone's attention during a banquet in his honor, when he declared, in what two fellow Norwegians characterized as his "highly original English":

> I tank youse very mooch and dees is all vary nice and youse vary
> kind to mes, but I ask youse ven I am here vy don't youse take
> dese vales at your doors—dems vary big vales and I seen dem
> in houndreds and tousends.

After his adventures in the *Jason* and the whaling voyages of the *Grönland,* the *Antarctic,* and the Dundee fleet, it had taken this extraordinary expedition to bring the prospect of commercial whaling in the Antarctic to fruition. And this time around, in sharp contrast to the snakebit ventures of the past, the cards fell in place perfectly.

For one thing, the Argentine coast and vicinity actually was home to some of the right whales the earlier expeditions had vainly sought. For another, Larsen had identified an enclosed harbor on South Georgia as the perfect site for a shore station. And although most of the whales in that area were rorquals, a good proportion of these were humpbacks, which are slower and swim nearer to shore than the rest of their close relatives. Finally, even the faster rorquals were no longer out of reach: even as the British had been writing off the prospect of whaling for rorquals in perpetuity, Svend Foyn had developed a deck-mounted harpoon cannon, which enabled whalers to shoot the faster whales from a ship rather than from small boats and retain them without sinking.

Larsen's comments soon attracted interest from a number of expatriate investors in Argentina, and in February 1904, a matter of weeks after Larsen's return to civilization, the Compañía Argentina de Pesca SA was constituted in Buenos Aires. On November 16 of that year, its first catcher was anchored in what Larsen called Grytviken, or Cauldron Bay—after the try-pots in which whale blubber was rendered—and on December 22 the operation shot its first whale, a humpback. About two months later, the first 990 barrels of whale oil produced in the Antarctic were shipped back to Argentina. Over its first twelve months, the company shot 183 whales—149 humpback, 16 fin, 11 blue, and 7 right whales, netting 5,488 barrels of oil, which sold for $69,886, and $13,372 worth of baleen.

In subsequent years, other companies set up shop on South Georgia, leading to a slaughter that, in the words of whaling historians J. N. Tønnessen and A. O. Johnsen, "surpassed anything ever seen in the history of whaling." In 1910, the amount of whale oil brought out of South Georgia exceeded the entire global total of the previous three years. In the first decade of South Georgia operations, whalers killed a recorded 1,738 blue whales, 4,776 fin whales, and 21,894 humpback whales—in other words, more humpbacks than are now believed to be alive throughout the world. The slaughter of the humpbacks was as effortless as it was merciless: gunners reported that they could go right up to the whales and simply drop their harpoons on them. In addition, they found that if they killed one of a pair of whales, its mate would not leave its side and thus also offered itself as easy prey.

Inevitably, in a number of years, humpbacks had all but disappeared from the waters around South Georgia. No matter: in an almost precise replication of the spread of sealers before them, whaling operations became established in the South Shetland and South Orkney Islands. In January 1906, the first of the so-called factory ships arrived in the region—the Norwegian vessel *Admiralen,* which anchored off Deception Island. As the generic name implies, these ships were floating factories that processed and cut up whales at sea, allowing the fleets to extend their reach and kill more whales in less time: instead of having to head back to the shore station after killing only a few whales, the catcher boats could travel farther, stay out longer, and simply deposit their catch before returning to the hunt. The original factory vessels were limited in their operation, however; the space was inadequate to allow for the whole whale to be processed, and the equipment was insufficient to handle the larger rorquals. As humpbacks became scarce and regulations were passed requiring full utilization of whales that were killed, use of the original floating factories began to fall away. It was, once again, Carl Anton Larsen who brought factory ship whaling in the Southern Ocean into a new era.

In November 1923, Larsen led a converted steamer from Hobart, Tasmania, en route to the Ross Sea. At 8,223 gross tons, the *Sir James Clark Ross* was the largest whaling ship at that point in history. It was outfitted with cookers, pressure boilers, capacity for 58,000 barrels of oil, and accommodations for a crew of 170, and it towed five whale catchers to its destination.

Its maiden voyage was, however, in the words of writer Richard Ellis, "a qualified failure." The fleet did not take very many whales, and many of those killed could not be processed at sea. Despite its great size, the *Sir James Clark Ross* was hampered by many of the same restrictions that had limited the smaller floating factories. It had to anchor in a harbor, and even this giant ship could not easily handle the big blue and fin whales. Nonetheless, it was a sign of things to come, and within two years, the final piece of the puzzle would be put in place: the development of an invention by gunner Peter Sørlle of a stern slipway up which dead whales could be winched for flensing and processing on deck. In 1925, the first factory ship so equipped, the *Lancing,* left Sandefjord, a seaport in Norway, for the Antarctic. On the way, it paused to take 294 humpbacks in West African waters, every one of them hauled up and flensed on deck. The *Lancing* was the way of the future; thereafter, giant factory ships were developed that could be truly independent, staying at sea for months on end if necessary, taking whales from and providing fuel for their own fleet of catcher vessels. Now no part of the ocean was outside the whalers' reach.

Larsen would not live to see this final development. He died on the *Sir James Clark Ross,* and his body, per his wishes, was returned to Sandefjord for burial. But the juggernaut he had helped set in motion could not be stopped. Commercial whaling, an industry that had been in existence for several hundred years, was about to enter its final phase—a phase that would ultimately prove self-destructive, but not until the industry had slashed almost every population of whales in the Southern Ocean to the smallest fraction of its original numbers.

Growth was rapid. The British had arrived shortly after the Norwegians; for most of the period of large-scale Antarctic whaling, these two nations would dominate the catch. But other countries and republics—among them, at various stages, Argentina, Denmark, Germany, Japan, the Netherlands, Panama, South Africa, the Soviet Union, and the United States—also joined the fray. Whale oil was being put to a wide variety of uses, from margarine to gunpowder, and the potential for profits was alluring. By 1930, thirty-eight floating factories with stern slipways were plying the Southern Ocean, accompanied by 184 catcher boats. In the 1919–1920 pelagic whaling season, 5,441 whales were killed in the Antarctic. By 1927–1928, this figure had grown to 13,775. Just two seasons later, the catch had almost tripled again, to 40,201. All told, between 1900 and 1940, a recorded 510,000 whales were killed in Antarctic waters.

There were some efforts at regulation, initially prompted primarily by Norway. In 1931, an international convention was drafted that was based heavily on a 1929 piece of Norwegian domestic legislation, the Act on the Taking of Baleen Whales. The killing of right whales was prohibited under this agreement, as was the taking of calves and females with young. Mention in the agreement of measures by which these regulations might be enforced, however, was conspicuous in its absence, and not until 1935 did the convention gather enough signatures to enter into effect. Two years later, the whaling nations— minus Japan, which refused to attend—met in London to strengthen the 1931 agreement. Minimum size limits were set for different whale species; whales below those limits were supposed to be immune from hunting. All pelagic (open-ocean) whaling was banned between the equator and 40°S latitude. Rules were adopted requiring full utilization of the whale carcass. Seasons were established during which hunting could take place and outside of which it could not. Inspectors were to be placed on board vessels. The following year, additional measures were added: the humpback was granted full protection in the Antarctic (a moot move by that stage, given that the humpback had been hunted to near-extinction in the region) and a part of the Southern Ocean was set aside as a "sanctuary" where no whaling could take place.

The agreement had little immediate effect: barely had the ink dried than South Africa began whaling in the Antarctic, one month before the season was scheduled to open, on December 8. In 1937–1938, the first season after the improved convention was signed, 46,039 whales were recorded as being killed in the Antarctic, the highest total yet.

World War II did not completely end whaling in the Antarctic, although it did result in a significant reduction as ships and manpower were assigned to other tasks. After the war, in December 1946, representatives of fifteen whaling countries gathered in Washington, D.C., to sign an entirely new regulatory agreement, the International Convention for the Regulation of Whaling. Three years later, the International Whaling Commission (IWC) met for the first time.

The International Whaling Commission

The preamble to the 1946 convention starts with encouraging words. It recognizes "the interest of the nations of the world in safeguarding for future generations the great natural resources represented by the whale stocks." It notes that "the history of whaling has seen over-fishing of one area after another and of one species of whale after another to such a degree that it is essential to protect all species of whales from further over-fishing." But the inherent contradictions that were to plague the IWC become clear toward the end of the preamble. The purpose of the convention, it concludes, is "to provide for the proper conservation of whale stocks" and at the same time, seemingly in direct contradiction, "thus make possible the orderly development of the whaling industry." In attempting to achieve both of those goals, the IWC, at least for the first twenty to thirty years of its existence, completely failed to meet either.

The fundamental problem was that the IWC's members—and therefore the world's whaling nations—were essentially regulating and policing themselves. Although there was a Scientific Committee that advised the IWC on catch levels, this advice was all too often ignored—unless it could be interpreted as meaning that a particular whale population was larger than had previously been considered. It only took one meeting, for example, for the IWC to decide that humpback stocks in the Southern Ocean had recovered and to lift the ban on hunting them in the Antarctic; it was another fourteen years before the ban was reinstated.

Individual whaling nations frequently took advantage of a clause in the convention—by which any country that did not like an IWC decision could lodge an official objection and thus not be bound to it—to cajole their fellow members into agreeing to larger quotas. Or they could threaten that if they did not get their way, they would just pick up their footballs and leave. For example, in 1959, in the face of protests from several countries over proposed quota reductions, the report of that year's IWC meeting noted: "Conscious of the importance of maintaining the Convention, the Commission showed a willingness to consider making some increase in the Antarctic catch if thereby the loss of those member countries which had given notice of withdrawal could be averted." The effectiveness of the withdrawal threat thus underlined, the Netherlands—which had been lobbying for less restrictive limits—walked out of the IWC, and it stayed out until 1962.

Perhaps the most egregious error of the IWC's first twenty-some years, however, was the means by which quotas were set: not on a species-by-species or country-by-country basis but through the Blue Whale Unit, or BWU. The BWU was meant to reflect the amount of oil that could be secured from one whale of any of several different species. One BWU was equivalent to one blue whale, two fin whales, two and a half humpback whales, or six sei whales. The first quota established by the IWC was 16,000 BWU—an arbitrary figure but one that could perhaps be seen as some kind of progress, given that in the last full season prior to World War II, about 30,000 had been netted. The consequence of its formalization as a way of setting quotas, however, was to hasten the very overexploitation the IWC had been founded to prevent. Killing a blue whale was just as easy as killing a fin whale—and now officially twice as lucrative. Accordingly, the immediate effect of the IWC's coming into being was the continued slaughter of the Antarctic's blue whales.

The blue whale is the largest creature ever to live on this planet. It can reach 100 feet in length and weigh as much as 160 tons. Even a newborn blue whale is 25 feet long. An adult blue whale is so big that an African elephant, the largest land mammal, could fit on its tongue. But these same remarkable qualities proved to be the blue whale's undoing. Once whalers were able to hunt rorquals, they found all that oil and meat in one package too good a target to resist. Whaling fleets targeted it with abandon: 1929–1930 saw 17,898 blues killed in the Antarctic; the following year, 29,410—the highest number ever. The blue whale catch had already begun to drop before the outbreak of war; the BWU served to punctuate the slaughter. In 1949–1950, the first season after the IWC's first meeting, the recorded catch was 6,182. The following year, it rose above 7,000 for the last time. By the 1955–1956 season, the catch had dropped to 1,614; by 1962–1963, it was 947.

As early as 1954, the IWC's Scientific Committee urged that blue whales be protected in at least part of the Antarctic; the IWC denied the request, although it did note that "it may soon become necessary to restrict more severely the Antarctic catch of blue whales." The following year, the Scientific Committee concluded that the previous season's catch figures indicated further declines in the blue whale population in the Southern Ocean, and it expressed concern that the whale's "powers of recovery might already be found to be largely lost, even if it received total protection." In 1960, the IWC appointed an independent panel of scientists to determine the level of kill each species could sustain without its existence coming under threat. Three years later, this "Committee of Three" reported that there should be a complete ban on the hunting of blue whales in the Antarctic and of humpbacks throughout the Southern Hemisphere. If such a ban were immediately imposed, they said, stocks of these whales would still need fifty years to recover. In response, the IWC did indeed reintroduce the ban on humpback whaling, and it also banned the killing of blue whales through much of the Antarctic. This latter move, though, was disingenuous at best: the area that remained open to blue whale hunting was the same part of the Antarctic that had yielded virtually the entire blue whale catch in recent years.

The following year, 1964, the IWC underlined the fact that any protective measures it passed were the exception rather than the rule when it proved incapable of putting any limit at all on the Antarctic catch. And then, in the 1964–1965 season, the once-unthinkable happened. All the whalers operating in the Southern Ocean were able to find just twenty blue whales. They killed them all. At its next meeting, the IWC finally granted the blue whale complete protection. By then, it was too late. The species was all but extinct in the Antarctic.

It was more than eighty years after James Clark Ross reported his sightings of "a very great number of the largest-sized black whales" before the factory ship that Carl Anton Larsen named after the explorer killed the first blue whale in the sea that also bears his name. It took only forty years more for the blue whale to be all but wiped out from the Southern Hemisphere. This is, by any reckoning, an appalling legacy.

Tens of thousands of whales have died since the blue whale was belatedly declared off limits. But by then, the heyday of Antarctic whaling was over. Before the 1960s were out, diminished catches had prompted almost all the major whaling nations to hang up their harpoons, leaving Japan and the Soviet Union to divide the Antarctic spoils between them.

As the blue whale declined, catches of the next largest species, the fin whale, increased: from 9,697 in 1935–1936 to 14,381 in 1936–1937 and up to 28,009 in 1937–1938. Throughout the 1950s, the fin whale catch was consistently in the high 20,000s, and then, suddenly, it also collapsed: from 27,099 in 1961–1962 to 18,668 in 1962–1963, 14,422 in 1963–1964, and 7,811 in 1964–1965. Ultimately, the fin whale would also be protected in the Antarctic, although the whalers' immediate response was sanguine: they simply switched their attention to the next on the list, the sei whale. In 1962–1963, the official recorded catch of sei whales in the Antarctic was 5,503; two years later, with the collapse of the fin whale hunt, the number of seis recorded as killed soared to 20,380. By the end of the decade, the sei catch was also in decline, and by the early 1970s, the whalers were concentrating their efforts on the minke whale—the smallest of the rorquals and previously considered too small to be worth hunting. It has subsequently been disclosed that even when the IWC did assign protection to whale populations, the Soviet Union rendered that protection meaningless. In 1961–1962, for example, the factory ship *Sovetskya Rossiya* killed 1,568 humpbacks, although the entire Soviet fleet reported just 270. Between World War II and 1972, the Soviets killed almost 48,500 humpbacks, of which they reported only 2,700. They also killed 3,212 right whales but admitted to killing just one.

In 1972, the IWC finally abandoned the BWU. That same year, representatives of countries attending the United Nations Conference on the Human Environment in Stockholm voted overwhelmingly for a ten-year moratorium on commercial whaling. Although efforts to replicate that success within the IWC over the next two years met with failure, the tide was turning. The long, haunting "songs" of the humpback whale were recorded by researchers and, improbably, released as a best-selling album. They were considered so beautiful that they were included on the *Sounds of Earth* recording carried by the *Voyager* spacecraft on its journey beyond our solar system. Other scientists noted the size of whale and dolphin brains and speculated on cetaceans' possible intelligence. Although some of these theories—such as the belief of one researcher that dolphins had their own language, "Delphinese," and would one day merit a seat at the United Nations—pushed the envelope of scientific credibility, they did add to a growing sense that the slaughter of the world's whales was morally wrong. As a more general environmental ethic began to take root, the burgeoning "save the whales" movement became its poster child. Beginning in the mid-1970s, Greenpeace took the fight against commercial whaling to the front lines, and footage of its activists steering inflatable boats between whales and harpoons was beamed via television into living rooms worldwide.

Eventually, the IWC caught up to public opinion. In 1982, it agreed to an indefinite global moratorium on commercial whaling, to take effect as of the 1985–1986 Antarctic whaling season. In 1994, underlining the particular importance of the Antarctic and the extreme extent of the devastation wreaked there, the IWC added the extra security of declaring virtually all waters south of 40°S latitude to be a Southern Ocean Sanctuary, free from commercial whaling even if the moratorium were to be lifted in the future.

The story is not entirely over: as of this writing (October 2000), Japan continues to ignore both sanctuary and moratorium, killing approximately 400 minke whales each year in the Antarctic for what it claims is "scientific research." There are concerns, too, about the prospects of future whaling on a larger scale, for which Japan in particular continues to lobby. But the days in which fleets of catchers and giant factory ships scoured the Southern Ocean and killed thousands upon thousands of whales are almost certainly finished. And although the belated campaigns against whaling deserve much credit for the slaughter's ending when it did, the sad truth is that in the Antarctic, at least, the whales simply are no longer there. Only minke whales, the last Antarctic species to have been targeted, remain at anything like pre-exploitation levels, and in 2000 the IWC's Scientific Committee reported that apparently even the minkes were significantly less numerous than had been presumed. It is doubtful that Antarctica will ever again see the numbers of whales at which Ross so eloquently marveled.

Much has been written about the killing of whales, in the Antarctic and elsewhere, and about the consequences of the commercial whaling industry's rampant disregard for the results of its own actions. Among the commentaries is one succinct but poignant observation by R. B. Robertson, who traveled to South Georgia in 1950 as ship's surgeon aboard a British factory ship. Four years later, he published a book in which he described the scene that met him there—the appalling stench and sight of men, covered with blood and grime, slicing their way through whale carcass after whale carcass.

It was, he wrote, "the most sordid, unsanitary habitation of white men to be found the world over, and the most nauseating example of what commercial greed can do at the expense of human dignity. I think that, if Captain Cook were to see it today, he would probably burst into tears."

C h a p t e r   5

The Last Wilderness

It would be a number of decades before commercial whaling faded entirely from South Georgia. Although several of the island's whaling stations had closed by the 1930s, four of them—Grytviken, Leith, Stromness, and Husvik—struggled on until the 1960s, when they, too, conceded to the inevitable. At its peak, whaling on South Georgia employed close to six thousand men; today, there are only hints of the bustling industry that once dominated these shores. Buildings stand derelict and defaced with graffiti. Oil drums and other wreckage lie scattered across the ground. Whaling ships sit at the slipway or wallow, half beached, on the shore.

In 1984, British Broadcasting Corporation journalist Robert Fox described the whaling stations on South Georgia as looking

> as if the steamers appeared in the bay one day, the factory whistle blew and everyone packed for home. At Stromness a railway wagon fully loaded with equipment has stopped halfway through the door of the main shed. In one of the accommodation huts a carton of tinned food has been left half opened. Leith is stacked with lengths of iron and piping, never used, and the storage bins are piled with hundreds of plumbing spares, washers, nuts and bolts.

117

The whalers were the only ones ever to really settle South Georgia, but in the decades following whaling's demise, the stations were visited occasionally by tourists and curiosity seekers and more regularly by crews of long-range fishing vessels. Many of the latter left the old whaling stations in even worse states of disrepair than they found them, helping themselves to any potentially useful hardware they could find and leaving behind tins, bottles, and trash.

By 1981, Leith in particular was such an eyesore that an effort was made to clean it up. The Christian Salvesen Company, which held the lease for the station, put out tenders for scrap merchants who could rid the place of the worst-offending junk, and the contract was won by a Buenos Aires–based firm, Georgias del Sur SA, run by Señor Constantin Davidoff. He applied to the British government, which administered South Georgia, for permission to visit the island and inspect the station—as well as those at Husvik and Stromness, which his contract also covered—and to the Argentine Ministry of Foreign Affairs for use of one of that country's naval vessels as transport. Neither request was in itself particularly unusual, and both were granted.

Protocol required that anyone visiting South Georgia officially inform the designated "magistrate"—the term given to the head of the British Antarctic Survey (BAS) scientific base at King Edward Point, Grytviken—and secure his permission to land. Davidoff, however, made no such approach. After leaving Buenos Aires on December 16, 1981, his transport ship—the *Almirante Irizar,* a naval icebreaker of Argentina's Antarctic Squadron—maintained radio silence throughout its five-day journey. On arrival, it bypassed Grytviken and headed straight for Stromness Bay, site of the three other whaling stations. By the time BAS discovered that the visit had been made, Davidoff and the *Almirante Irizar* had already left. The magistrate visited Leith on December 23 and found signs of the Argentine visit, including, scrawled on a wall, the words *Las Malvinas son Argentinas:* The Malvinas are Argentine. *Malvinas* is the name by which the Spanish-speaking world refers to the Falkland Islands, a pair of subantarctic islands about 900 miles west-northwest of South Georgia; the message could in hindsight be considered a presage of things to come. But Argentina had long expressed sovereignty over the islands, and the British, though disconcerted by the happenings at Leith, chose not to initiate any formal protest, determining that it "would risk provoking a most serious incident which could escalate and have an unforeseeable outcome."

A little more than two months later, Davidoff was back on South Georgia. He had apologized profusely to British officials for any trouble he might have caused and expressed his desire to return to the island with his men to begin cleanup work at the whaling stations. Britain raised no objections, and so on March 9, Davidoff formally informed the British embassy that he and forty-one employees were about to set sail with the Argentine navy for South Georgia, where they planned to stay and work for about four months. Two days later they departed on board the *Bahia Buen Suceso;* just prior to their leaving, David-off's lawyer telephoned the embassy once more to confirm the departure and was reminded of the vessel's obligation to report to King Edward Point on arrival.

But, as had the *Almirante Irizar* before it, the *Bahia Buen Suceso* ignored the required formalities and sailed straight to Leith, where it began disembarking crew and unloading equipment. On March 19, a party of BAS personnel discovered the ship at anchor and came across about 100 people wandering around ashore. Some of them had been shooting the island's reindeer. High on a pole, fluttering in the breeze, flew the Argentine flag.

Within a couple of days the ship had left, but this time not everybody it had brought to the island departed with it. About forty people stayed behind, and after discussions back and forth, the British informed the Argentines that those remaining were now considered an occupation force and unless they were retrieved by Argentina, steps would be taken to remove them.

In response to the British announcement, Argentina dispatched a naval auxiliary vessel, the *Bahia Paraiso,* to South Georgia, where it disembarked a complement of marines. The evolving crisis was debated in the House of Commons in London. The British press picked up the scent: ARGENTINE INVASION OF SOUTH GEORGIA ISLAND, exclaimed the *London Evening Standard.* The Argentine fleet started to assemble off the country's coast. Prime Minister Margaret Thatcher initiated preparations for Britain to dispatch a naval contingent of its own. And then, on April 2, 1982, Argentina invaded the Falkland Islands. The following day, after a fight of several hours, the small number of Royal Marines on South

Georgia was overwhelmed, and South Georgia also fell. A British task force, comprising a total of more than 100 ships, headed south to take the islands back. War was about to reach the edge of the Antarctic Regions.

✳ ✳ ✳

There is no certainty as to who saw the Falkland Islands first. Argentine histories accord the honor to various Portuguese and Spanish seamen, particularly Esteban Gómez, a member of Ferdinand Magellan's expedition of 1522; in the English-speaking world, credit is normally given to English navigator John Davis on board the *Desire* seventy years later. Neither claim has much substantiation. The first undisputed sighting belongs to Dutchman Sebald de Weerdt, sometime around 1600. None of these three tarried around his "discovery," however, and no human foot that we know of touched the islands until 1690. That foot belonged to John Strong, an English sailor who was on a voyage to Chile when blown off course by a storm, which forced him toward the islands' northern coast. He was not exactly fulsome in his praise for the place: there was, he noted, "fresh water in plenty" and an "abundance of geese and ducks," but "as for wood," he wrote, "there is none." Strong charted the sound between the two islands, named it after the first lord of the Admiralty, Viscount Falkland, and continued on his way.

Despite such an uninviting summary of the islands' worth, in the eighteenth century covetous eyes looked at them longingly. Britain's Lord Anson, considering them possible vital staging posts for trade routes around Cape Horn, opined that "even in time of peace," they "might be of great consequence to this nation and in time of war would make us master of the seas." Anson was unaware—or, just as likely, did not care—that theoretically the question of the islands' sovereignty had already been settled by the Treaty of Utrecht, signed in 1713 as part of a wider-ranging peace among a number of European powers. Among the provisions of the treaty were that Spain ceded Gibraltar to the British; France did likewise with its North American territories of Newfoundland, Nova Scotia, and Hudson Bay; and all parties reaffirmed Spain's dominion over its traditional territories in the Americas—including such outlying islands as the Falklands.

If the treaty did not deter the British from giving voice to their interest in the islands, it also apparently meant little to the French. In 1764, seeking in some obscure way to gain revenge for his country's loss of Quebec to England, French nobleman Louis-Antoine de Bougainville landed on East Falkland Island, claimed both islands in the name of King Louis XV, and established a colony that he and his men called Port Louis.

The following year, Commodore John "Foul-weather Jack" Byron arrived on West Falkland, hoisted the Union Jack, and, usefully, planted a vegetable patch before sailing away. The next summer, a few hardy souls returned to take advantage of Byron's gardening skills and set up a small settlement of their own.

Spain was furious. Both Byron and de Bougainville, the Spanish pointed out, had clearly violated the Treaty of Utrecht, an insult that could not go unanswered. France swiftly sought to ease the tension by hammering out an agreement whereby, in exchange for reparations to de Bougainville, they would hand their colony over to the Spanish. To those Spaniards who actually took possession of the islands, this diplomatic victory was decidedly Pyrrhic: "I tarry on this miserable desert, suffering everything for the love of God," lamented Father Sebastian Villeneuva, the first priest at what was now known as the colony of Puerto Soledad. Nonetheless, the fact that the early Spaniards disliked the islands did not mean they were inclined to leave them to the British, and in 1769, five ships and 1,400 men set out from Buenos Aires with instructions to rid the Falklands of any remaining Englishmen.

Rather than providing an effective end to the dispute, that action served merely to stoke the fires of nationalist sentiment in Britain and ignited a popular movement to establish a permanent British presence on the islands— much to the discomfort of the government, which had little interest in doing anything of the kind. So anxious, in fact, was the government to dampen its countrymen's ardor that it commissioned Samuel Johnson to belittle the islands' importance and desirability. Dr. Johnson, whose view of civilization's extent pretty much ended at the municipal boundaries of London, needed little persuasion. West Falkland, he wrote, was "thrown aside from human use, stormy in winter, barren in summer, an island which not even the southern savages have dignified with habitation, where a garrison must be kept in a state that contemplates with envy the exiles of Siberia, of which the expense will be perpetual."

Eventually, a settlement was reached whereby Spain, though not in any way conceding sovereignty, agreed to allow a small British party to return to West Falkland "to restore the King's honor." But that offer was conditional on the British agreeing to leave again afterward—a condition the two governments kept secret for fear of further fanning the flames of public protest in England.

The British duly returned, installed a plaque proclaiming British sovereignty, stayed for three years, and left. In 1790, Britain and Spain signed the Nootka Sound Convention, in which Britain formally renounced territorial ambitions in South America "and the islands adjacent." Spain maintained an uninterrupted presence at Puerto Soledad until, in 1811, it began to lose control over its South American colonies and removed all its settlers from both the islands and the Patagonian mainland.

In 1816, the forerunner of the modern Argentine republic formally declared independence from Spain; four years later it claimed the Falklands, by dint of the right to postcolonial succession. Since the departure of the Spanish, the Falklands had descended into anarchy, becoming a kind of subantarctic Wild West where sealing vessels sought refuge from storms and what passed for a rule of law was applied by vessel captains. Accordingly, in 1829, Buenos Aires appointed a governor, Louis Vernet, to impose order and facilitate the establishment of a new settlement. Vernet's task was far from easy, with many of the sealers vehemently hostile to the imposition of any outside authority. When, in 1831, he attempted to enforce Argentine fishing regulations by arresting three U.S. sealing vessels, he touched off a crisis that resulted in the loss of Argentina's grip on the islands.

In response to Vernet's action, Silas Duncan, captain of the USS *Lexington*—which happened to be in Río de la Plata at the time—steamed to the Falklands, landed, destroyed all military installations, razed all the buildings, and then sailed away, declaring the islands free of government. Argentina sent a second governor to restore order, but meanwhile the British saw an opportunity. They sent the sloop *Clio* to the Falklands, where its captain, James Onslow, calmly informed the Argentines that the British were asserting sovereignty and that the Union Jack would be raised the following day, January 3, 1833. Outnumbered and outgunned, the Argentines retreated to the mainland without firing a shot. Despite Argentine protests, Britain formally established the Falkland Islands as a crown colony in 1840. Over the years, successive governments in Argentina seethed with anger and resentment until finally, in 1982, a military junta, using the islands as a diversion from political and economic turmoil at home, launched an invasion.

For all the history building up to it, when war finally erupted it was relatively brief. Argentine forces on South Georgia surrendered almost as soon as British soldiers stepped ashore on April 25, 1982; the main battle for the Falklands lasted seventy-two days, culminating in the capture of the capital, Port Stanley, on June 14. But despite its brevity, it was brutal: Argentine fighters sank several British ships, and nearly 900 people were killed—236 British and 655 Argentine, many of the latter ill-prepared conscripts hurriedly drafted by the junta.

International responses to the conflict varied. Most members of the Organization of American States closed ranks around Argentina, supporting its campaign against what it saw as British colonialism. Much of the European Community and the British Commonwealth supported the United Kingdom—as did, after a period of attempting to broker a peace, the United States. Whatever the official positions of their governments, however, many people around the world viewed the battle with puzzlement and some incredulity. It seemed remarkable that in the penultimate decade of the twentieth century, two developed countries could go to war over what U.S. president Ronald Reagan referred to as "that little ice-cold bunch of land down there," a seemingly worthless outcrop of rock on the fringes of the Antarctic. Even to some of the combatants, the conflict had a surreal quality: in their book *The Battle for the Falklands,* Max Hastings and Simon Jenkins, characterizing the war as a "freak of history," note that "even after men began to die, many of those taking part felt as if they had been swept away into fantasy, that the ships sinking and the guns firing around them had somehow escaped from a television screen in the living room."

Yet for all the battle's freakish qualities and the seemingly unfathomable motivations for going to war over a piece of subantarctic real estate, Britain and Argentina were hardly alone in having designs on territory in the region. Indeed, some critics of the war argued that the conflict was not really about the Falklands and South Georgia at all but had to do with establishing access to the real prize: the continent of Antarctica itself. Throughout the twentieth century, nation after nation had claimed Antarctic territory, and now, even as the world looked askance while Britain and Argentina squared off, a great many more countries were seeking to secure a piece of the Antarctic pie.

The Final Frontier

The ball had been set in motion three-quarters of a century earlier by the British. Citing the explorations and discoveries of James Clark Ross, Robert Falcon Scott, Edward Bransfield, William Smith, and others, Britain in 1908 formally laid claim to the land and islands lying south of 50°S and between 20°W and 80°W, a claim that encompassed the Antarctic Peninsula region and extended far enough north to include the Falkland Islands and South Georgia. Unfortunately, in achieving the latter it also claimed for London the southern cone of Chile and Argentina, and so in 1917 the claim was revised slightly, keeping essentially the same longitudinal coordinates but circumventing South America.

Following World War I, Britain sought to expand its reach. In a 1920 memorandum to the governors-general of Australia and New Zealand, Leopold Amery, British under-secretary of state for colonies, stated:

> HMG [His Majesty's Government] have under consideration the future policy of the Empire in the Antarctic regions, and in particular the question of the control of the mainland and of the neighbouring waters . . . with the exception of Chile and Argentina and a few barren islands belonging to France, every inhabited land in the direction of the Antarctic regions is already British. . . . HMG have, therefore, come to the conclusion that it is desirable that the whole of the Antarctic should ultimately be included in the British Empire.

However, the memo stated, "the time has not yet arrived that a claim to all the continental territories should be put forward publicly." The British plan was to advance in stages, taking advantage of the Antarctic proximity of Australia and New Zealand to lodge claims those colonies could administer. The first step came on July 30, 1923, when the British government, with the approval of the

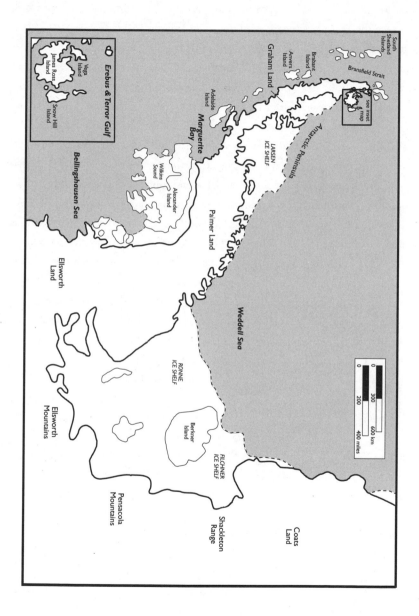

South
Shetland
Islands

Bransfield Strait

see inset
map

Graham Land

Anvers
Island

Brabant
Island

Adelaide
Island

LARSEN
ICE SHELF

Antarctic Peninsula

Marguerite
Bay

Wilkins
Sound

Alexander
Island

Palmer Land

Bellingshausen Sea

Ellsworth
Land

Weddell Sea

Ellsworth
Mountains

RONNE
ICE SHELF

Berkner
Island

FILCHNER
ICE SHELF

Pensacola
Mountains

Shackleton
Range

Coats
Land

Erebus & Terror Gulf

Vega
Island

James Ross
Island

Snow Hill
Island

0
0
200   400 miles
300   600 km

government of New Zealand, declared that "that part of His Majesty's Dominions in the Antarctic Seas which comprises all the islands and territories
between the 160th degree of east Longitude and the 150th degree of West Longitude which are situated south of the 60th degree of South latitude, shall be
named the Ross Dependency." The governor-general of New Zealand was
vested with authority "to make all such Rules and Regulations as may lawfully
be made for the peace, order and good government of the said Dependency."

Ten years later, in February 1933, the Australian Antarctic Territory (AAT)
was declared as being that area south of 60°S and between 160°E and 45°E.
The territory in question was enormous, covering approximately 2.4 million
square miles (excluding ice shelves) and accounting for approximately 40 percent of the continental landmass. It would have been even larger except that it
was divided in two by Terre Adélie (Adélie Land), a small slice that had been
claimed by France in 1924. There had been some opposition within British and
Australian circles to recognizing the French claim; in 1928, it had been argued
that "the preference of the British government and even more of the Dominions was that the Empire should have no neighbours at all in the Antarctic or in
its adjacent islands." But considerations of realpolitik ultimately won the day,
with Britain determining that the French claim—which had been asserted in
response to that of New Zealand—implied acknowledgment of the legitimacy
of the Ross Dependency and that mutual recognition would facilitate the AAT's
adoption.

Meanwhile, Norway had grown sufficiently concerned that the British
might try to claim the entire Antarctic coastline, and thus perhaps attempt to
exclude Norway from whaling in the Southern Ocean, that it had made a pre-
emptive claim to Bouvet Island in 1928, which it followed up by claiming Peter
I Island in 1931. Finally, in 1939—its hand forced by interest in the region on
the part of Nazi Germany—it formally claimed the area immediately to the
west of the AAT, a region that had been extensively mapped and charted by an
expedition under the patronage of Lars Christensen and that Norway dubbed
Queen Maud Land.

The growing encroachment of foreign powers on the continent ultimately
prompted action from the two countries nearest to it, Chile and Argentina. In
November 1940, Chilean president Aguirre Cerda declared the establishment

of a Chilean Antarctic Territory, "formed by all lands, islands, islets, reefs, glaciers and pack ice, and others, known or unknown, and the respective territorial sea existing within the segment constituted by the meridians 53 degrees longitude west of Greenwich and 90 degrees longitude east of Greenwich." Such a move, argued Canas Montalva, chief of the southern military region, was essential because "Chile must call attention to herself, beside the other powers that claim ownership over the southern polar hemisphere, precisely fixing the logical boundaries of her proprietorship, even though her modest resources have not permitted this until now, establishing the sovereignty that was necessary or giving it the commercial exploitation that it offers."

Argentine president Julio Roca had announced his country's intention to establish new bases in the "southern seas of the republic" considerably earlier in the twentieth century. In 1913, representatives of the republic talked with the British about the latter ceding the South Orkney Islands on the grounds that Argentina had continuously occupied a weather station on one of them, Laurie Island, since 1904. The talks went nowhere, but Argentina stayed, adding a post office, hoisting the national flag over the station, and scattering plaques declaring Argentine sovereignty across the island. During World War II, with the attentions of the British elsewhere, Argentine expeditions occupied and took over British bases in the Antarctic Peninsula region without hindrance, depositing claims to sovereignty wherever they went. In 1943, Argentina joined Britain, New Zealand, Australia, France, Norway, and Chile and became the seventh country to lay formal claim to Antarctic territory.

The following year, the British ships HMS *William Scoresby* and SS *Fitzroy* sailed into the harbor at Deception Island in the South Shetland Islands as part of an expedition named Operation Tabarin. In part, Tabarin was formed in response to concerns about Nazi Germany's use of Antarctic and subantarctic islands as havens for its warships. In 1940, the U-boat *Pinguin* had found and destroyed an entire fleet of Norwegian whalers off Queen Maud Land; although the *Pinguin* was sunk by HMS *Cornwall* in 1941, its sister vessels the *Komet* and *Atlantis* had continued to cruise the Southern Ocean and launch raids on Allied ships farther north. In addition, the Nazi interest in Antarctica had been highlighted by a 1939 expedition in which two hydroplanes, launched by catapult from the deck of the *Schwabenland,* dropped five-foot-long aluminum

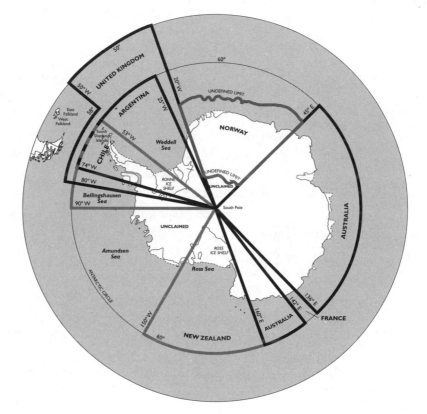

darts, the Nazi swastika engraved on their tails, over large parts of Queen Maud Land. The height from which they were dropped allowed the darts to penetrate solid ice; presumably, many are even now buried in the ice shelf. The Nazis' plan was to use them as markers by which to delineate the Antarctic territory they intended to claim.

In 1941, HMS *Queen of Bermuda* had visited the old whaling station at Deception Island, destroyed the remaining stocks of coal, and blown holes in the fuel oil tanks, thus denying them to any Nazi ships that might appear there. Operation Tabarin sought to follow up such efforts, but it also was a mission to underscore the British presence in the region. Not without purpose, Chile's Antarctic claim overlapped Britain's, and Argentina's took up part of both. Each was seeking to establish its own authority at the expense of the others, and initially, at least, the Chileans and Argentines were less concerned with each other than with presenting a united Latin front against the British. The British, in turn, were particularly anxious not to cede any ground to an Argentine government sympathetic to the Axis powers, and when the *Scoresby* and *Fitzroy* sailed into Deception, their primary goal was to rid the island of any signs of an Argentine presence. A 1942 expedition in the *Primero de Mayo* had visited Deception Island and left sovereignty claims; in 1943, HMS *Carnarvon Castle* had obliterated these sovereignty marks, only for another Argentine expedition to put them up again. Now it was the turn of the ships of Operation Tabarin. Notes the official report:

> For a time several of the landing party were actively engaged in the removal or obliteration of sundry items of evidence of disrespect shown to lawful British sovereignty by unfriendly visitors to the island. The halyards of the flag-pole erected by one of HM ships the previous year had been removed. However, N.F. Layther, the base radio officer, volunteered to reeve new halyards, and after a climb rendered difficult by malicious mutilation of the flag-pole, he succeeded in doing so and the Union Jack was broken at the top by 4.30 PM on 3rd February. This simple but significant act of allegiance was loyally acclaimed by all those assisting at the ceremony.

Argentina protested the British action but this time did not return the favor. Nonetheless, right up until the time of the Falklands conflict, tensions between the two continued in Antarctica. In 1948, Britain sent warships to protest Argentine and Chilean "incursions"; in 1952, Argentine soldiers fired over the heads of a British party attempting to land at Hope Bay on the Antarctic Peninsula; in 1976, a group of Argentine "technicians" took possession of Southern Thule, an island in the South Shetlands; that same year, an Argentine warship intercepted the BAS ship *Shackleton* south of the Falklands and, when the British ignored an order to stop because it was "in Argentine waters," fired across the bow.

Despite their differences, the seven claimant states were united on one issue. However much Britain contested Argentina's claim to the Antarctic or Norway bristled at the British assumption of entitlement to the entire continent, all were in general agreement that they did not want anyone else laying claim to Antarctica. Each of the seven had lodged its claim either to make a case for sovereignty over all or part of the continent or to protect its self-interest from those of the others. The more nations that took a bite out of the pie, the smaller the influence and status of each of the seven would be.

Unfortunately for them, however, in the years immediately preceding and following World War II, one other country had begun showing a definite interest in Antarctica. Furthermore, this country emerged from the conflict more powerful than ever before and swiftly positioned itself as the most important player in the Antarctic. That country was the United States.

* * *

American sealers had raced to South Georgia almost as soon as Captain James Cook returned to England in 1775. As a consequence, noted historian Kenneth Bertrand, American interest in the Antarctic was "as old as the nation itself. . . . There has been no decade when at least some Americans have not been in the Antarctic as seal hunters, whalers, or explorers." Seventy-five years before Carsten Borchgrevink—or one of his comrades—hopped ashore on the coast of the Ross Sea, American sealing captain John Davis may well have been the first man to set foot on the continent, at Hughes Bay on the Antarctic Peninsula. Antarctica was once held by many—and still is by some—to have been discovered by an American, Nathaniel Palmer. The name accorded to the peninsula's southern part, Palmer Land, is in his honor, although Palmer's logbooks suggest he saw the continent ten months after Bellingshausen, Bransfield, and Smith.

The first official U.S. voyage of exploration into the region, the United States Exploring Expedition of 1838–1842, visited Antarctic waters as one component of an impressive, ambitious global journey of discovery and scientific research that also visited such disparate areas as the Atlantic Ocean, Brazil, Tierra del Fuego, Chile, the Pacific Ocean, Australia, New Zealand, the western coast of North America, the Philippines, and the East Indies. Led by Lieutenant Charles Wilkes, the expedition's fleet was in the Antarctic for only sixtynine days, but during that time it accomplished a great deal, surveying approximately 1,600 miles of coastline and providing the first firm evidence that Antarctica was a continent.

The voyage was not without its difficulties: one ship was sent home for being too slow, another was lost at sea, and during the course of the whole expedition, sixty-two men, chafing against Wilkes' tyrannical qualities, were discharged, and forty-two others deserted. In addition, the reports Wilkes made of his Antarctic findings on his return to the United States were cast into doubt by James Clark Ross, who pointed out that he had sailed across areas Wilkes had marked as being mountainous landmasses. It was some time before it was fully appreciated that both men were right. Wilkes had indeed seen the mountains he charted, but he had been fooled by the phenomenon of fata morgana, or polar refraction: the clear, dry polar air caused a mirage similar to those seen in deserts wherein distant objects seem closer than they really are. But such controversies, combined with a series of charges related to his treatment of his crews, meant that Wilkes, instead of being praised for his efforts, was roundly castigated and summarily court-martialed.

Despite the controversy surrounding Wilkes himself, the expedition as a whole was a success. Kenneth Bertrand referred to it as "a milestone in American science [which] had a marked effect on the scientific activity of the young republic." The expedition brought back "such a wealth of geological, botanical, zoological, anthropological, and other materials that they served as a foundation of much of American science, and it indirectly exerted a major influence on the form of organization of the Smithsonian Institution." Nonetheless, for the rest of the nineteenth century, the United States—and, for that matter, most nations involved in the region—more or less left Antarctica to the sealers. It would be the best part of a hundred years after Wilkes before American expeditions returned to the Antarctic in any great number.

The so-called heroic age of Antarctic exploration, which had been dominated by Europeans and Australians, came to an unofficial end with the death of one of its greatest figures, Ernest Shackleton, at South Georgia in 1922. By the second quarter of the twentieth century it had given way to a new era, which has been called the mechanical age, and it was the dawn of this era that truly marked the appearance of the United States on the Antarctic stage. U.S. explorers were swifter than most to capitalize on the advantages of new technology in combating hostile terrain; in particular, they took full advantage of the arrival of aviation.

Between 1933 and 1939, Lincoln Ellsworth, son of a Pennsylvania mining millionaire, collaborated with Australian Hubert Wilkins on several Antarctic expeditions with the goal of making an airplane flight across the continent. Frustrated by damage to his plane on the first attempt and foiled by severe weather on the second, he succeeded on the third, flying with Herbert Hollick-Kenyon in three stages before running out of fuel and being forced to land on the Ross Ice Shelf. In 1947, Finn Ronne, who would become one of the great names in Antarctic exploration, launched a major expedition that included a series of flights over the Antarctic Peninsula region. As a result of these flights, Ronne was able to determine that the Ross and Weddell Seas did not meet, as had earlier been postulated, and thus did not slice Antarctica in half; in the process, he went a long way toward confirming that Antarctica is in fact a continental landmass.* But the man who more than any other prompted and

---

*At the surface level, anyway. Subsequent studies have shown that beneath the massive ice cap, the situation is a little different. The bulk of the continent, East Antarctica, is one big, cohesive landmass; West Antarctica—which, broadly speaking, encompasses the Antarctic Peninsula and the Weddell Sea west to the Ross Sea—is of more recent geologic origin. The two halves came together as a result of plate tectonics. The Transantarctic Mountains, which rim the western coast of the Ross Sea, are considered the dividing line between them.

encouraged U.S. interest and involvement in Antarctica was undoubtedly Admiral Richard Byrd.

Byrd had already claimed to be the first to fly to the North Pole and had been the second person to fly solo across the Atlantic when, on November 28, 1929—Thanksgiving Day—he flew with Bernt Balchen, Ashley McKinley, and Harold June to the South Pole and back, accomplishing in sixteen hours a round trip that Roald Amundsen's team had taken three months to complete and that Scott and his men could not finish. Five years later, during a second Antarctic expedition, Byrd spent several months alone in a hut on the Ross Ice Shelf before being rescued as he slowly succumbed to fumes from his kerosene stove.

Almost lost in the aura of such headline-grabbing highlights, Byrd's expeditions added a great deal to knowledge of the Antarctic continent. The first voyage included a winter sledding expedition south along the Ross Ice Shelf to the Queen Maud Mountains; a flight over what, in honor of the admiral's wife, became known as Marie Byrd Land, east of the ice shelf; and the discovery in Marie Byrd Land of a mountain range, which Byrd called the Edsel Ford Ranges and is now officially referred to as the Ford Ranges. Members of the second expedition roughly delineated the eastern edge of the Ross Sea, mapped and christened several previously unseen mountains, and discovered an island, which they named Roosevelt Island, covered by ice and surrounded by the Ross Ice Shelf. On rocks halfway up Scott Glacier they found lichens, the most southerly plants yet found; nearby, coal-bearing formations on Mount Weaver, at Scott Glacier's head, yielded the southernmost known fossils. They identified bacteria, molds, and algae; studied bird life at sea, in the field, and at the Bay of Whales; and made a comprehensive study of the life history of the Weddell seal. All day, every day for 360 days, they conducted meteorological observations at the surface and in the upper atmosphere.

Combined with his previous aviation adventures, Byrd's Antarctic exploits made him a national hero. On his return to New York City after his flight to the South Pole, he became the first person in history to be rewarded with a second ticker-tape parade (the first had been for his transatlantic flight). The National Geographic Society's president, Gilbert Grosvenor, presented him with ten bound volumes containing five thousand letters from schoolchildren across the country. A letter to the *New York Times* argued that Byrd's accomplishments merited a newly detected planet being named after him. (The planet's discoverers demurred and called it Pluto.)

Previously, if the thoughts of Americans had been directed to a polar region at all, they had been focused on the Arctic and on the claims and counterclaims of Robert Peary and Frederick Cook, each of whom had insisted in 1909 that he had been first to the North Pole and that the other was a liar. (The evidence now suggests that they both faked their accomplishments.) Byrd, however, drew public attention to Antarctica, sparking a new fascination that was noticed at the highest levels.

In 1939, President Franklin Delano Roosevelt authorized an official, government-funded expedition to Antarctica—the first since Charles Wilkes' voyage slightly more than a century before. The United States Antarctic Service (USAS) expedition sought to map the coastline of Marie Byrd Land from the Antarctic Peninsula to the Ross Sea and to take the first steps toward establishing a permanent, year-round U.S. presence in the Antarctic—to prove, in Roosevelt's words, "that human beings can permanently occupy a portion of the Continent winter and summer." Plans for future expeditions were interrupted by World War II, but the year after the conflict ended, the United States returned to Antarctica in the form of Operation High Jump, the largest expedition ever to visit Antarctica—before or since. Its thirteen ships carried 4,700 personnel—more, probably, than had to that point visited the southern continent in total—as well as eight seaplanes, six R4D transport planes (more familiarly known in the civilian world as DC-3s), six helicopters, eight Army three-quarter-ton amphibian cargo carriers, ten Caterpillar tractors, Cletrac tractors, jeeps, and landing vehicles.

On a total of sixty-four flights about 70,000 photographs were taken, recording 60 percent of the Antarctic coastline, of which roughly 25 percent had never been seen before. All told, the expedition sighted some 1.5 million square miles of Antarctica, nearly half of it previously unexplored. Two planes, one of them carrying Richard Byrd, made the second flight to the South Pole.

Other nations with a stake in the region looked on in nervous apprehension, concerned that the USAS and Operation High Jump might presage the entry of the United States into their thus far exclusive, if sporadically dysfunctional, family. After all, one slice of the Antarctic remained unclaimed: the region west from the Antarctic Peninsula to the eastern Ross Sea. That region, Marie Byrd Land, was precisely the area that had been most extensively explored by the Americans; if the United States were to make a claim there, it would have a very good case. There was also a realization among the British, Chileans, and Argentines that the United States would, furthermore, have decent grounds for lodging a claim in the peninsula region. Even though Nathaniel Palmer's logbook showed him "discovering" the peninsula almost a year after Bransfield and Smith, at least his logbook was extant, unlike those of the British. And certainly, U.S. involvement in the region since that time, particularly in the parade of sealers who had made their way south in the 1800s and then through the USAS and Operation High Jump, matched Britain's and far exceeded those of either of the Latin American countries.

The official U.S. position, as articulated by Secretary of State Charles Evans Hughes in 1924, was that "the discovery of lands unknown to civilization, even when coupled with a formal taking of possession, does not support a valid claim of sovereignty unless the discovery is followed by an actual settlement of the discovered territory." Officially, therefore, the United States did not recognize any of the claims already in place, but it was fully prepared to make such a claim itself—indeed, one of High Jump's goals was to "consolidate and extend the basis for United States claims in the Antarctic, if such should subsequently be made."

On August 30, 1938, Secretary of State Cordell Hull cabled to Lincoln Ellsworth advice on how he could assert U.S. sovereignty over previously unseen and unexplored territory in Antarctica. Accordingly, during his flight across Antarctica on January 11, 1939, Ellsworth dropped a brass cylinder containing a message that claimed the area "for my country, so far as this act allows." Byrd also claimed areas for the United States, as had others before him, including Palmer and Davis. Furthermore, U.S. Department of State memorandums have come to light that show that even though it was not officially admitted, a large part of the rationale behind the USAS was "establishing and strengthening US claims within the sector previously explored by Admiral Byrd and Lincoln Ellsworth." Although no member of the USAS was to "take any action or make any statements tending to compromise the non-recognition position," members were nonetheless encouraged to "take any appropriate steps such as dropping written claims from airplanes, depositing such writings in cairns, et cetera, which might assist in supporting a sovereignty claim by the United States government." Secretly, also, Operation High Jump and subsequent expeditions were instructed to extend and consolidate "United States sovereignty over the largest practicable area of the Antarctic continent."

The United States had no interest in antagonizing those allies that already had a stake in the region, however. Prior to the departure of the USAS expedition, President Roosevelt even floated the notion of giving "some study to a new form of sovereignty, i.e., a claim to sovereignty of the whole sector lying south of the Americas in behalf of, and in trust for, the American Republics as whole." The idea never gained any traction, and Chile and Argentina shortly thereafter filed claims of their own.

As it turned out, the view steadily evolved in Washington that the country's desired aims could be achieved just as easily, or even more so, without claiming territory. Part of the rationale for Britain's acknowledging Adélie Land, after all, was that the French claim implicitly legitimized the Ross Dependency; similarly, had the United States claimed, say, Marie Byrd Land, this could have been taken as an admission that, for example, New Zealand had authority over the Ross Sea sector or Norway over Queen Maud Land. By not playing that game, the United States would be free to operate wherever it wanted in Antarctica.

There were reasons why the United States did not want to place any restrictions on its Antarctic operations. In the eyes of Pentagon planners, the vast, barely explored wilderness presented an enormous vista that was perfect for training troops and testing matériel for battle. With East and West abutting each other at the other end of the world, military leaders assumed that were World War III to break out, at least some of the fighting would be in the Arctic. Although there had been some military exercises in the north—Operation Nanook in Alaska and Operation Frostbite in Davis Strait—they had not been conducted far enough north and had not, for sundry political reasons, been as big as the navy had wanted them to be. Part of the rationale for Operation High Jump was to allow the navy to flex its muscles and use Antarctica as a training ground for polar warfare, a training ground to which many in the Pentagon sought to have permanent, unfettered access.

As much as keeping all of Antarctica freely available for its use in exercising its own interests, the United States was motivated by the desire to keep its enemies out. The Nazis' claim to what they called *Neu Schwabenland* and their success in using Antarctic and subantarctic harbors as bases for their submarine raiders had shown the potential value of Antarctica to a hostile regime. Even though that threat had been disposed of, a new foe was making its presence felt around the world.

From the time of Bellingshausen onward, Russia and its successor state had paid little attention to Antarctica beyond objecting to Norway's claim of Peter I Island, on the grounds that it had been discovered by Bellingshausen's expedition—that it had indeed been the very first land sighted south of the Antarctic Circle. But following World War II, the USSR, perhaps in response to U.S. efforts, had begun to assert interests of its own. The country's position mirrored that of the United States: it did not and would not claim territory in Antarctica, but it reserved the right to do so. Nonetheless, the Soviets clearly intended to make their presence felt, and in 1949, the Geographical Society of the Soviet Union adopted a three-part resolution asserting Russian priority in the discovery of Antarctica and noting the "indisputable right of the Soviet Union to participate in the solution of problems of the Antarctic."

Prior to that declaration, the initial U.S. response to rumblings from the Kremlin had been the diplomatic equivalent of putting its fingers in its ears and humming loudly, as manifested in a 1948 proposal that the United States and the seven claimant nations merge their interests and administer the Antarctic as a form of condominium. This offended some other countries that had begun reminding the rest of the world of their own interests in the Antarctic: Belgium, for example, which had launched an expedition to the Antarctic Peninsula in 1897, and South Africa, which cited its annexation of subantarctic Marion Island as well as its relative proximity to the continent. But the United States feared that opening up the proposal to others besides itself and the claimant states would only make it harder to justify keeping out the Soviet Union.

It did not matter; the proposal went nowhere. There was a seemingly unbridgeable chasm between, on the one hand, the United States—which refused to acknowledge any claims—and the other nations, which mostly insisted on the legitimacy and irrevocability of their own claims. There were some differences of opinion among the seven: New Zealand expressed a willingness to forgo a claim if a workable international regime could be developed but preferred it to be under the United Nations' auspices, and Britain nibbled tentatively at the bait, interested as it was in a resolution to its territorial conflicts in the Antarctic Peninsula region. But France was noncommittal, Australia and Norway saw no need for international involvement, and Argentina and Chile were adamantly opposed. Besides, even if the eight had managed to work things out among them, the USSR was not about to throw up its hands and admit defeat. The Soviets made it clear that they expected to be an integral part of any further decision making regarding Antarctica's status and that any attempt by the United States and the claimant states to make a backroom deal that excluded them would only lead to trouble.

Shortly thereafter, North Korea invaded South Korea, the United States became distracted, and the prospect of any rapprochement between East and West on any territorial issue appeared doomed. It began to seem likely that even Antarctica—the forbidding wilderness at the end of the Earth—was about to become yet another front line between political and nationalist ideologies, the coldest battleground of the cold war.

And then, suddenly, unexpectedly, the stars aligned themselves perfectly, and at a stroke Antarctica's status changed—from cold war frontier to a model of international cooperation. It was a development that owed much to circumstances in the world beyond, but it was set in motion by an ambitious program of scientific study, and it would ultimately lead to what has been widely declared one of the most successful international treaties ever written.

"Antarctica Shall Be Used for Peaceful Purposes Only"

It began in suburban Maryland on April 5, 1950, during a dinner party at the home of physicist James Van Allen. In a few years, the sun would be approaching its periodic solar maximum—an approximately eleven-year cycle, poorly understood to this day, at the end of which the star's electromagnetic activity peaks. For amateur astronomers, the cycle is notable because of the increased number of sunspots on the solar surface; for observers in the polar regions, it results in spectacular displays of aurora borealis and aurora australis—the northern and southern lights, the result of solar particles interacting with Earth's magnetic poles and exciting the ionized gases in the upper levels of the atmosphere. For researchers, it opens a window to learning more about the star at the heart of our solar system. To let slip such an excellent opportunity to study the causes and effects of these solar dynamics would be tragic, opined the assembled dinner guests. What could be done to maximize the scientific potential?

Among those at the party was Lloyd Berkner, a scientist who had served as a radio technician on Byrd's first expedition to Antarctica and who had later played a role in the development of radar; for some months he had been considering the notion of an International Polar Year to focus and consolidate research on the Arctic and Antarctic. The first such event had taken place in 1882–1883, under the auspices of the International Polar Commission. It had been restricted almost entirely to Arctic studies, and ground-level studies at that; similarly, although the Second International Polar Year, held fifty years later, had been planned to include a ring of bases around Antarctica, the effects of the Great Depression had killed that idea and, once again, the Arctic had been the sole beneficiary. Staying true to the spacing between the previous polar years would have resulted in the third one taking place in 1982. Instead, offered Van Allen, how about halving the interval, which would mean the next one would come around in 1957–1958—exactly the time of the solar maximum?

The idea took root, and in 1952 the International Council of Scientific Unions (ICSU) adopted a formalized version and appointed a special coordinating committee. The World Meteorological Organization chimed in, urging that the effort's scope be expanded beyond the polar areas, and representatives of other sciences argued that their disciplines should also be involved; thus the Third International Polar Year became the International Geophysical Year (IGY). To take greatest advantage of the solar maximum and to provide the most opportunity for data collection, the "year" would last from July 1, 1957, to December 31, 1958. Its focus would be on two realms whose mysteries science had only begun to uncover: outer space and Antarctica.

The scientific community was resolute in its determination that the undertaking should be untainted by political considerations. Proposals for research programs should be submitted and reviewed on the basis of their scientific merit alone. Such noble intentions would, of course, have counted for nothing were it not for the support accorded the venture by several governments, including those of the United States and the USSR. Soviet scientists were keen to participate, and the Kremlin welcomed the IGY as an opportunity to become involved in Antarctica without having to endure hostility from other interested nations. Meanwhile, the administration of U.S. president Dwight D. Eisenhower, seeing the chance to further consolidate the position of the United States as the dominant actor in the Antarctic theater, pledged a massive infusion of funds and matériel.

Perhaps the single most important decision was made early in the process: that any bases established in the course of the IGY would be for scientific observation only and would not denote any political significance or territorial aspirations. Countries were largely free to set up bases in those areas they thought were of greatest scientific value, without having to worry about the sensitivities of claimant states. There was an occasional exception: Australia was less receptive to the notion than were most of the other claimants, and when it bristled at having to accept a Soviet station in its territory, Belgium and Japan, which both wanted bases nearby, were encouraged to pick other sites to avoid disturbing the peace.

Political considerations also came into play in a quasi-veto of a Soviet plan to set up shop at the most symbolic and significant of all Antarctic locations. At a 1955 planning meeting in Paris for the IGY, the Soviet delegation, which arrived late, announced that the USSR would establish three Antarctic stations, including one at the South Pole, but Georges Laclavère, chairman of the coordinating committee, swiftly countered that the United States had already committed to its own base at the Pole and that to have two would be repetitious. There was, however, another pole still up for grabs, one that would be even more demanding and that, unlike the geographic South Pole, had never before been reached: the Pole of Relative Inaccessibility, so called because it is the farthest point inland from all coasts. Very well, agreed the Russians, "we do not insist on the geographic pole."

As it happened, the United States had at that stage made no such commitment, as Laclavère knew well, although the proposal had been put to the Eisenhower administration. Fortunately, however, Washington agreed to the suggestion, and so, on October 31, 1956, the U.S. Navy R4D Skytrain aircraft *Que Sera Sera* landed at the South Pole and six men stepped out onto the polar plateau. The team's leader, Admiral George Dufek, had been a part of both the USAS expedition and Operation High Jump; during the former, he had stood with Richard Byrd as the two looked out over the vast Ross Ice Shelf.

"What," asked Byrd, "do you think of it?"

"It's a hell of a lot of ice," Dufek replied, "but what use is it?"

Despite such misgivings, Dufek had become an important player in U.S. Antarctic efforts and was ultimately appointed tactical leader of Operation Deepfreeze, the U.S. component of the IGY. Now, forty-five years after Robert Falcon Scott and his four comrades turned around and disconsolately began their ultimately fatal journey north, he was the third party leader, the eleventh person overall and the first American, to stand at the bottom of the planet.

Within minutes, Dufek and his team fell victim to frostbite, the effect of −58°F temperatures heightened by a strong breeze, which was itself accentuated by the blast of the plane's propellers. The men were able to dig a hole deep enough to plant a flag, but Dufek's efforts at taking photographs met with failure: his cameras were frozen. When one of the crew said that he could no longer move his fingers and that they had done everything they could, Dufek replied: "Good. Let's get the hell out of here."

When the frostbitten Dufek returned from his brief mission, he declared conditions in the Antarctic too terrible to contemplate building a base there. By November, however, the weather had improved, and on November 20, 1956, construction began. Suddenly, this most remote and forbidding of areas was buzzing with activity, with planes dropping supplies, and tractors busily fashioning a base beneath the ice. Paul Siple, who had traveled to Antarctica as a Boy Scout on Byrd's first expedition, placed a silvered glass ball atop an orange-and-black-striped bamboo pole: a symbolic pole at the Pole. On December 15, almost a quarter of a million letters that had been sent by stamp collectors were laboriously franked by hand with POLE STATION, ANTARCTICA, and five days later, at midnight on the summer solstice, the base was declared open for business. The winter of 1957–1958 saw men spend the winter at the South Pole for the first time in history; the bottom of the world has been permanently inhabited ever since.

The inaugural overwintering party had barely settled in when they welcomed their first visitors. Just four years previously, Edmund Hillary had, with Tenzing Norgay, been the first atop Mount Everest; on January 4, 1958, he and four fellow New Zealanders reached the South Pole, the first people to do so by land since Scott. As they stood there, a team led by Englishman Vivian Fuchs was making its way toward the Pole from the Weddell Sea. Guided by Hillary and supplied by depots the New Zealanders had laid for them, they would continue on to the Ross Sea, becoming the first people to traverse Antarctica by land, and when Fuchs arrived at the New Zealand base on McMurdo Sound on March 2, 1958, a knighthood was waiting for him.

Although not officially part of the scientific studies that constituted the IGY, the British Commonwealth Trans-Antarctic Expedition was conducted in tandem with them, and it became perhaps the most visible and celebrated feat of that long year. But the IGY's legacy would last long after Fuchs and Hillary returned home to receive their merited plaudits. What had been conceived almost on a whim as a worthy scientific venture during a dinner party had unexpectedly transformed the way governments regarded Antarctica.

The IGY had been a phenomenal success, not only in the amount of scientific data generated but also in the extent of international cooperation it permitted in Antarctica. Prior to the IGY, for example, relations between Australia and the USSR were at a temperature that would have made ice-cold seem balmy, largely as the result of a spy scandal, but despite initial Australian concerns, the Soviets occupied a base at Mirnyi in the Australian Antarctic Territory, and the two nations exchanged data and logistic support.

At the same time the IGY showed that such close cooperation was possible, it also made it all the more desirable: the Soviet Union now had its sought-after toehold on the frozen continent, and it could not and would not be dislodged. Fears remained that unless an effort was swiftly made to capitalize on the momentum of friendship and camaraderie fostered by the IGY, the worst nightmares of the West in the previous decade or so might yet come to pass. Noted *New York Times* journalist Cy Sulzberger in a dispatch from the South Pole:

> It is in our interest to insure that this vast region should never be turned by the Russians into a kind of Antarctic Albania. Moscow now threatens Greece and Turkey from rocket ramps in that Adriatic land. Similar launching pads, operated by a handful of men, could be used by the U.S.S.R. to blackmail the entire Southern Hemisphere from here.

Similarly, a 1959 article in the magazine *Missile and Rockets* had voiced official concerns in dramatic tones:

> At the frozen bottom of the earth Russia is moving into a position from which its missile squadrons could outflank the free world. Half of Antarctica is rapidly turning from white to red. . . . From the snowy Antarctic coasts and plateau Russia is in a position to have the entire Southern Hemisphere within easy missile range. . . . The Soviets can put their [intercontinental ballistic missiles] on Antarctica itself. . . . South America and Africa would become much more sensitive to Soviet pressure as the shadow of Soviet missiles might spread across both continents. . . . Meanwhile, Soviet tractors bearing red flags are moving through the Antarctic snows in increasing numbers.

But even as such hostile rhetoric belied the success of the IGY, cooler heads were elsewhere prevailing. Britain had been consulting quietly with its Commonwealth allies with an eye to revisiting the earlier failed talks on internationalization of Antarctica, and in a press conference in Australia in February 1957, British prime minister Harold Macmillan proposed two major principles for settlement of the Antarctic question: that there should be "free trade in science" and that Antarctica "should not be allowed to develop" into an area "used for military purposes."

At about the same time, the United States also was reexamining what to do politically with Antarctica. On May 3, 1958, a note from Washington was delivered to the eleven other nations that had actively participated in the IGY— Argentina, Australia, Belgium, Chile, France, Japan, New Zealand, Norway, South Africa, the United Kingdom, and the USSR—proposing that "the interest of mankind would best be served, in consonance with the high ideals of the Charter of United Nations, if the countries which have a direct interest in Antarctica were to join together in the conclusion of a treaty."

Although Argentina and Chile were hesitant, response to the U.S. note was generally favorable; after almost eighteen months of discussions, negotiations, and meetings of working groups, representatives of the twelve nations gathered in Washington, D.C., on October 15, 1959, to hammer out a treaty. By December 1, the Antarctic Treaty was ready for signature, and on June 23, 1961, after it had been ratified by the governments of all twelve signatories, it came into force.

✳ ✳ ✳

Article I of the Antarctic Treaty states that "Antarctica shall be used for peaceful purposes only. There shall be prohibited, inter alia, any measures of a military nature, such as the establishment of military bases and fortifications, the carrying out of military maneuvers, as well as the testing of any type of weapons." Specifically, Article V underlines that any "nuclear explosions in Antarctica and the disposal there of radioactive waste material shall be prohibited." In order to assuage the fears of hawks that no matter how fine-sounding the treaty, one of the signatories could agree to such provisions and then blatantly ignore them, Article VII provides the right for any contracting party to "designate observers to carry out any inspection" of any other country's bases, with such observers having "complete freedom of access at any time to any or all areas of Antarctica."

Coupled with such freedom-of-access provisions, perhaps the single most important aspect of the treaty—certainly the one that was integral to its acceptance and adoption by the claimant states—is Article IV. While specifying that nothing in the treaty should be considered as renouncing, diminishing, or prejudicing any existing claim to territory in the Antarctica, it makes explicit the following:

> No acts or activities taking place while the present treaty is in force shall constitute a basis for asserting, supporting or denying a claim to territorial sovereignty in Antarctica or create any rights of sovereignty in Antarctica. No new claim, or enlargement of an existing claim, to territorial sovereignty in Antarctica shall be asserted while the present Treaty is in force.

In other words, the treaty essentially dealt with the controversy of sovereignty by putting it to one side—by freezing future claims and papering over those already in existence. It allowed the seven claimants to continue asserting their rights, removed from their concerns the possibility of other nations challenging those rights, and ensured that countries other than those seven had the freedom to operate in Antarctica as and where they wished, regardless of territorial claims.

The treaty's primary purpose, as stated in its preamble, is to ensure "in the interests of all mankind that Antarctica shall continue forever to be used exclusively for peaceful purposes and shall not become the scene or object of international discord." Despite their various earlier differences, its member states have shown a degree of cooperation and support remarkable for an international organization; for example, even as they went to war in 1982, Britain and Argentina sat down together at treaty meetings as if nothing were untoward. Initially meeting approximately every two years, and eventually annually, treaty nations steadily increased in number: Czechoslovakia joined in 1962, Denmark in 1965, the Netherlands in 1967. Although full "consultative" status was bestowed only on members that demonstrated an established presence in the Antarctic, this privileged level of membership ultimately grew to include such disparate countries as Poland, Brazil, India, China, Finland, and Peru.

In 1964, treaty members inked the first round of a series of Agreed Measures for the Conservation of Antarctic Fauna and Flora, which forbade killing, capturing, or interfering with native mammals or birds without a permit; obliged members to minimize harmful interference with the Antarctic environment and to limit water pollution; enacted a series of Specially Protected Areas (SPAs) and Sites of Special Scientific Interest (SSSIs); and prohibited importation of nonindigenous species except under permit. In 1972, they

adopted the Convention for the Conservation of Antarctic Seals; the convention, which came into force in 1978, banned the killing of some seal species
while severely restricting the possible hunting of others, essentially ending
modern commercial sealing in the Antarctic before it began. In 1982, treaty
members added the Convention on the Conservation of Antarctic Marine Living Resources (CCAMLR), which was developed in response to overexploitation of fish populations in the Southern Ocean and in anticipation of a possible
major commercial fishery for krill.

All these developments could be viewed as mileposts along a continuing
path toward the internationalization and conservation of Antarctica. But
national self-interest and Antarctica were not mutually exclusive just yet, and
one issue in particular remained to be resolved. From the outset, although
many had lauded the Antarctic Treaty as a model of international cooperation,
others had cautioned that this would be true only so long as such cooperation
was worth everyone's while: if Antarctica ever showed promise of profitability,
all those involved in the frozen continent would happily cast aside their decla-
rations of principle and scramble over one another in a contest to extract the
most riches. And if that were to happen, there would be near-unanimity over
its probable cause: in the reported words of one observer, the Antarctic Treaty
would "last till a big mineral discovery is made—then it will be every man for
himself."

✳ ✳ ✳

Take a look at the continents of the world and it will not be long before you
realize that they seem, many of them, as if they could be giant pieces in a crude
jigsaw puzzle. As far back as 1620, English philosopher Francis Bacon noted in
Novum Organum that eastern South America appears as if it could nestle comfortably into the western coast of Africa. But despite the obviousness of the fit
and the similarities of other coastlines elsewhere, the orthodoxy was that the
continents were rigid and unmoving and such appearances were little more
than charming coincidences. Not until 1915 did the German physicist Alfred
Wegener first propose a scientific theory to account for the phenomenon. In
The Origin of Continents and Oceans, he postulated the notion of "continental
drift," wherein all the continents had at one stage been joined in a gigantic land-

mass, which Wegener called Pangaea, surrounded by a super-ocean, Panthalassa. At the end of the Triassic period, a little more than 200 million years ago, he suggested, Pangaea began to break up and drift apart, eventually forming the continents we know today. Wegener did not rest his thesis solely on the jigsaw fit of continental coastlines; he noted that the fossils of two small prehistoric reptiles were found only in limited areas of western Africa and eastern South America and that coal deposits in Pennsylvania, France, and Siberia contain fossils of equatorial and tropical vegetation. He also noted similarities in the ages and rock layers of mountain ranges in Appalachia and parts of Europe as well as in the fossils found there.

Wegener's theory was not popularly accepted, and his cause was not helped by the fact that the mechanism he proposed to explain his theory—that tidal forces influenced by the moon's gravity pushed the continents around—was massively unconvincing. But if his specifics were lacking, his general thesis would ultimately be endorsed with the discovery in the 1960s that Earth's entire surface was composed of a series of "crustal blocks," or plates, which were in constant—if slow—motion in relation to one another.

Thus was born the science of plate tectonics, and a good many things on Earth were attributed to its effect, from mountain chains and rift valleys to earthquakes and volcanic eruptions. The notion that the continents were not always as we know them now, that they had split apart from one or more larger landmasses, was confirmed and explained, albeit with some refinements. There had indeed been a Pangaea and a Panthalassa, and when the giant landmass broke apart it did so primarily into two parts, both of them far larger than any contemporary continent: to the north, Laurasia, consisting primarily of what became North America and the bulk of Eurasia; and to the south, Gondwana, which ultimately split up into India, Madagascar, Africa, Australia, New Zealand, South America, and Antarctica.

The evidence that Antarctica had once been something other than a frozen, isolated continent had long been there to see. Fossils of ancient plants more suited to a tropical environment had been found on the South Shetland Islands almost immediately on the islands' discovery, and Carl Anton Larsen had uncovered fossils on Antarctica itself in 1892, as had Otto Nordenskjöld during his enforced stay on the continent, Edward Wilson during Scott's expedition, and numerous others since. In 1967, New Zealand geologist Peter Barrett found the first fossilized Antarctic land animal—the jawbone of a roughly 220-million-year-old lizard called a labyrinthodont. Subsequent fossil finds have included aquatic plesiosaurs, plant-eating hadrosaurs, *Cryolophosaurus ellioti* (a meat-eating dinosaur from the Jurassic period), and even an entire fossilized swamp.

Of more significance, though, than such remains was the prospect that if Antarctica had indeed been conjoined with other southern continents, it might well share some of their geologic traits. Perhaps Antarctica—along its coasts, or possibly beneath its icy mantle—harbored mineral riches like those of which some of its former neighbors boasted.

That Antarctica might contain minerals was not a new proposition. It had been assumed so from the start simply because every other continent has them. Successive waves of exploration have uncovered deposits of different substances in various amounts and of inconsistent quality: iron in East Antarctica, coal throughout the Transantarctic Mountains, sand and gravel scattered here and there, relatively small amounts of metals such as copper, zinc, lead, gold, platinum, and manganese. The issue of what to do about such minerals in the event that deposits should be found of sufficient quality and quantity to justify a commercial extraction industry came up during negotiations for the Antarctic Treaty but immediately proved far too nettlesome. It was one thing agreeing to freeze sovereignty claims and to allow freedom of access for scientific researchers and government-appointed inspectors. But the question of what to do and who would benefit in the event a commercial mining operation got well under way on Antarctica was another matter entirely. Because of the problems it presented and the risk it posed to the delicate emerging post-IGY consensus, treaty negotiators dealt with the minerals issue by leaving it out of the convention entirely.

That might not have mattered much had things stayed the way they were in 1959. But with the acceptance of the plate tectonics theory, thoughts turned to the possibility of Antarctica's mineral resources being richer than previously anticipated. Of particular interest was the notion of some commonality with the immense mineral wealth of South Africa, which claims a huge percentage of the world's diamonds, gold, and platinum, among others; and, perhaps above all, the prospect that oil and gas fields like those beneath the continental shelves of southeastern Australia, western New Zealand, eastern India, and Argentina might be located in the regions of the Ross and Weddell Seas.

Idle speculation morphed into genuine enthusiasm when, in 1973, the U.S. scientific drilling ship *Glomar Challenger* detected gaseous hydrocarbons in three holes drilled in the continental shelf of the Ross Sea. Although such emissions do not necessarily imply the presence of oil and gas deposits at further depths, they can do so, and when word leaked out that the U.S. Geological Survey (USGS) had estimated that 45 billion barrels of oil and 115 trillion cubic feet of natural gas lay beneath the Ross, Weddell, and Bellingshausen Seas, there was a sensation. Never mind the USGS' protests that these were estimates—very tentative estimates, later to be disavowed—of the *total* in-place deposits, that recoverable reserves were at best one-third of those totals, and that as a result, the amount of oil and gas beneath West Antarctica was roughly the same as that off the U.S. Atlantic coast and less than the total estimated reserves off Alaska. The genie had been let out of the bottle. At the same time, the major Arab countries exporting crude oil had severely curtailed supply, leading to fears of shortages and a sharp increase in fuel prices. The hunt was on for alternative sources, and Antarctica had just made its way onto the list.

Nonetheless, there was no sudden rush to drill for oil. First, there still was not (and still is not) solid evidence of large deposits in the region. Second, even if there were large deposits, the fact remained that Antarctica was a hellish place to go drilling and mining. The obstacles posed by foul weather, fearsomely cold temperatures, and giant icebergs dwarfed any problems associated with oil exploration in the Arctic. The cost of operating in such an appalling environment would be prohibitive; the technology was insufficiently advanced and the need for more oil insufficiently desperate to justify such an undertaking. And then there was the political component, a recognition that the Antarctic Treaty consensus would very likely crumble if a U.S. vessel definitively found oil in, say, the Ross Sea region and New Zealand asserted its right to the oil while the United States claimed discoverer's rights on behalf of its nationals. It would be better by far, it was agreed, to establish some kind of framework for such an eventuality before it actually happened. And so, at the Antarctic Treaty meeting in London in 1977, member states adopted a policy of "voluntary restraint" in which all those present agreed to "urge their nationals and other States to refrain from all exploration and exploitation of Antarctic mineral resources while making progress towards the timely adoption of an agreed regime concerning Antarctic mineral resource activities."

Such a regime first began to take shape in 1982; by early 1983, a draft treaty was circulated among treaty nations. Five years and six drafts later, the Convention on the Regulation of Antarctic Mineral Resource Activities (CRAMRA) was finished and ready for signing. Despite the effort invested in its development, however, CRAMRA never got off the ground.

From the beginning, even as the treaty nations began contemplating the concept of developing a regime for mineral regulation in Antarctica, those countries outside the Antarctic club were deeply disturbed by what they saw as a minerals bonanza being stitched up by a select few. Assurances that there would be no such windfall, that the minerals regime was being developed as a precautionary measure, fell on deaf ears. The critics, almost exclusively less developed countries, viewed the activities of the mostly industrialized Antarctic states with suspicion, believing that they were already exploring for oil under the guise of conducting scientific research.

Leading the charge was Malaysia, which in 1983 succeeded in having the question of Antarctica placed on the agenda of the United Nations General Assembly. Malaysia argued that what it perceived as the elitist, colonial framework of the Antarctic Treaty should be replaced by a kind of global commons, that Antarctica should be gathered into the warm embrace of United Nations trusteeship "for the benefit of mankind as a whole."

Although the Malaysians gathered a number of supporters to their cause, the effort stalled. A report by the United Nations secretary-general was bullish on the Antarctic Treaty's accomplishments, including maintaining peace and limiting adverse environmental effects. The treaty nations closed ranks and launched a diplomatic counteroffensive, pointing out among other things that membership in the treaty was explicitly open to any state in the United Nations. Although Malaysia and its allies enlisted the support of the Non-Aligned Movement, that effort was undercut by the readiness of some of the movement's largest and most influential members, such as India, to themselves enter the Antarctic Treaty fold.

By the end of the 1980s, the idea of Antarctica being replaced by a "global commons" for the economic benefit of all humanity had all but died. But so, too, had the notion of mineral exploration and extraction under the management of the treaty nations. The years between the *Glomar Challenger*'s drilling in the Ross Sea and the completion of CRAMRA had seen the rise throughout the industrialized Western world of a new ethic of environmental protection. An increasingly vocal and well-organized movement was protesting what it saw as continued, unchecked exploitation—indeed, overexploitation—of the world's natural resources, from whales to tigers to forests to fish. CRAMRA's supporters portrayed it as a conservation convention; they noted that under its provisions, "no Antarctic mineral resource activity will be permitted" unless it could be determined that it would not cause significant changes in the environment and its wildlife. In addition, all parties to CRAMRA would have to agree before any exploration was allowed.

For critics, however, that just was not enough. The language in CRAMRA, they argued, was vague and imprecise. What, for example, were "significant" changes? Elsewhere, the convention defined "damage to the Antarctic environment" as any adverse effect "beyond that which is negligible or which has been assessed and judged to be acceptable." But what was "negligible"? And who was to decide what was or was not "acceptable"?

However much CRAMRA sought to limit damage to the environment from mineral extraction, it accepted such extraction in principle—and that, argued critics, all but guaranteed that extraction would ultimately take place. If it did, they continued, all the rules, regulations, and safeguards in the world would not be enough to prevent environmental damage or forestall accidents such as oil spills—which in the fragile Antarctic environment could be devastating. Far better, they argued, to take a bolder step. Antarctica provided an opportunity—perhaps the last opportunity—to declare a part of the world off limits, to protect a sizable area of pristine wilderness before it was too late, to acknowledge that the fact that a resource *could* be exploited did not necessarily mean it *should* be exploited.

Founded by Washington, D.C., attorney Jim Barnes, the Antarctic and Southern Ocean Coalition (ASOC), an umbrella group of environmental organizations around the world, began lobbying for Antarctica to be declared a

"World Park," a notion that had been briefly floated by New Zealand in 1975. Barnes and others sought access to Antarctic Treaty meetings, testified before the United States Congress, and wrote impassioned articles. But despite their belief in the rightness of the cause and despite the sometimes high-profile support they attracted to it—including the support of such luminaries as marine explorer Jacques Cousteau, who organized a massive petition drive in support of the World Park campaign; Sir Peter Scott, son of the polar explorer and cofounder of the World Wildlife Fund; and Al Gore, at that time a senator from Tennessee—they could not derail CRAMRA. As a result, one of ASOC's member organizations decided that a more radical step was needed. The future and direction of the minerals regime was being dictated by the full, "consultative" members of the Antarctic Treaty. Consultative status was open to those who could demonstrate an ongoing presence in the Antarctic. Greenpeace elected to play the treaty nations' game: it would take a ship, go to Antarctica, and build a base. In so doing, it would declare its eligibility for a seat at the table. Although Greenpeace campaigners did not expect the organization's claim to be granted, they knew that such a step, though bold and risky, would help focus public attention like never before on the campaign to protect the frozen continent.

World Park Antarctica

Official response to the Greenpeace plan was less than enthusiastic. When the expedition ship MV *Greenpeace* made a port of call in Sydney on its first journey south in 1985, Australia's science minister, Barry Jones, descended with media in tow and declared the vessel unsuitable for polar travel. He had a point: even though the bow of the thirty-year-old tug had been strengthened for the task at hand, the ship had no ice rating and was vulnerable to Antarctic conditions. If it were to strike an iceberg or become trapped in the ice pack, the result could be calamitous. Not only would the ship's crew be in immense personal danger, but there was also the very real prospect of an oil spill that could cause the kind of damage to the Antarctic environment that Greenpeace sought to prevent. Even sympathetic environmentalists criticized Greenpeace's planning, fearing that the prospect of disaster could ruin any credibility the organization—and hence the World Park campaign—might have.

Outwardly, Greenpeace insisted that all necessary precautions had been taken and that no unnecessary risks would be; nonetheless, there were, within the organization and even on board the vessel, concerns about the venture's viability. This was, after all, the Antarctic: many a well-planned expedition had succumbed in some degree to its hostility.

The expeditioners cast such qualms aside and set out for their icy Grail, but even as they closed in on their destination, the naysayers' doubts seemed set to come to fruition. After anxiously skirting the edge of the pack ice in the Ross Sea, the *Greenpeace* received news that farther south, the *Southern Quest*—vessel of an expedition that sought to retrace Robert Falcon Scott's path to the South Pole—had been caught between ice floes and sunk. Unlike the *Greenpeace,* the *Quest* was an ice-class vessel. A short while later, unable to find a way through the ice to its destination, Ross Island, the *Greenpeace* turned around and headed for New Zealand.

The grand plan had been thwarted, but only temporarily. The *Greenpeace* returned to Antarctica the next year, and this time it reached its goal. On January 25, 1987, the ship entered McMurdo Sound; later that day, helicopters and inflatable boats made the first exploratory journeys to Cape Evans, along the beach from Scott's historic hut and the putative site of what Greenpeace planned to call World Park Base. By the end of the season, it was finished: a lurid green prefabricated hut emblazoned with the word *Greenpeace*. In 1988, the organization replaced the *Greenpeace* with the more appropriate ice-class MV *Gondwana,* and for several years the environmentalists were a permanent presence in the region—chasing Japanese whaling ships, recording the incidental capture of seabirds in lines set by Soviet fishing vessels, and, above all, making inspections of various countries' Antarctic bases.

A 1985 study by the Scientific Committee on Antarctic Research (SCAR), an arm of the Antarctic Treaty System, acknowledged that "the majority of existing research stations were established because these were the most convenient places for either logistical or scientific reasons and without thought for environmental effects." The construction of bases, noted the report, might require "radical modification" of the local habitat; oil spills and toxic wastes might be released into the sea; the resupplying of bases by ships and airplanes created the potential for accidental fuel spills; and the generation of power and heat might result in the emission of gases, waste heat, dust, and noise. In a series of reports presented to Antarctic Treaty nations, Greenpeace argued that the problems with bases went even further and that many of the member nations were not even paying lip service to the treaty's environmental provisions.

King George Island is the largest of the South Shetland Islands group, located just off the northwestern coast of the Antarctic Peninsula. In relation to much of the rest of the region, it is something of an oasis: less than 0.5 percent of the Antarctic coastline is free of ice year-round, and much of this is on King George Island and its environs. As a result, the area acted as a magnet for countries that wanted to build scientific bases in the Antarctic. The proliferation of such bases, claimed Greenpeace, in many instances had little if anything to do with the pursuit of science. Rather, because the best way to ensure full "consultative" status within the treaty was to demonstrate an established presence in the Antarctic, some countries were setting up bases purely as a way to get a seat at the table and a voice in negotiations on the minerals convention. With political concerns preeminent, in the mad rush to set up bases the Antarctic environment was sometimes trampled underfoot. On King George Island's Fildes Peninsula, the Russians built the Bellingshausen research station on land that had been designated a Specially Protected Area (SPA)—a region that under treaty guidelines is considered so biologically valuable that it should be subject to even less human activity than the rest of the continent. Once the SPA had been breached, other countries moved in. The peninsula's moss beds—rare beacons of green on a continent of gray and white—were severely damaged by vehicle tracks and construction work; a nearby lake was used for landfill; and the formerly pristine peninsula, argued Greenpeace, became "one of the most heavily polluted and damaged parts of Antarctica."

Longer-standing bases also came under criticism, with Greenpeace and others reserving particular opprobrium for the United States' McMurdo Station—located, like World Park Base, on Ross Island. McMurdo was and is by far the largest base operating in Antarctica, with as many as 1,200 people living and working there during summer. There was even, during the 1960s, a nuclear power plant to provide electricity for the base. The first and only such device in Antarctica, "Nukey-Poo" was removed, along with the contaminated ground on which it had stood, after a few years of operation. But even though the plant was gone, the sheer size of the base presented problems, not least in the amount of waste it generated. Solid waste dumps were set on fire several times each season, and other waste was discharged straight into McMurdo Sound. In common with those at several other bases, McMurdo personnel frequently left larger pieces of machinery, even disused vehicles, on the sea ice to sink slowly into the water during spring thaw. Today, whether because of public pressure—as Greenpeace claims—or, as U.S. officials insist, because they were planning on changing things anyway, McMurdo is much improved, and virtually all waste is now retrograded for disposal and destruction back in the United States.

Of all the targets for Greenpeace's criticism in the Antarctic, however, none generated quite as much heat and controversy as did the French station at Dumont d'Urville, in Adélie Land. It was not the existence of the station itself that was the problem but France's decision in the early 1980s to build a hard-rock airstrip to service it. According to the French, this was to enable the base to be resupplied by air rather than sea and to allow scientists to arrive earlier and leave later each season; as the French truthfully argued, sea ice in the Dumont d'Urville region was largely impenetrable for all but a couple of months each year, making resupply an arduous task and limiting the French to a shorter scientific study period than those of other countries. In response, a team of scientists, appointed by the French government to review the project as international criticism increased, asked why France could not simply do as the Americans did and use planes with skis on an ice runway. In the context of ongoing mineral negotiations, environmentalists viewed actions such as building new airstrips with particular suspicion. In this case, the suspicion was reinforced by the fact that contrary to Antarctic Treaty regulations, France had not conducted an environmental impact assessment or even informed its fellow treaty members of what it was doing.

To begin construction, the French dynamited five islands in the Pointe Géologie Archipelago, the plan being to level the islands' surfaces and use the dynamited material to fill in the channels between them. In so doing, however, they inflicted serious damage on the very bird life Dumont d'Urville had ostensibly been established to study. Unknown numbers of Adélie penguins and other birds were killed by the blasting, and as many as 8,000 birds of different species, which had previously used the islands as breeding grounds, were displaced. When news of the airstrip construction leaked in 1984, work was temporarily suspended, but by 1989 it had begun again.

Early that year, as construction workers looked on, boatloads of campaigners and crew members from MV *Gondwana* had shuttled ashore, carrying protest banners and signs. At first, the workers stood idly by and offered no resistance; but when they saw the prefabricated survival hut the protestors planned to erect in the middle of the runway, they leaped into their trucks and bulldozers and attempted to drive over it. In a scene more suited for student protests in Paris or Berkeley than the shores of Antarctica, Greenpeace activists scrambled into the hut, stood in front of the trucks, and threw themselves onto the blade of an oncoming bulldozer. Eventually, with tempers frayed all around, both sides agreed to a peaceful resolution to the protest. But by then, news of the confrontation—complete with videotape and photographs—had been broadcast and published around the world. Antarctica as an environmental issue had entered mainstream awareness as never before.

A few weeks later, another incident occurred that served to keep it there. On January 28, 1989, the *Bahia Paraiso,* now serving as an Argentine supply ship and tourist vessel, ran aground on rocks near the U.S. research station at Palmer on the Antarctic Peninsula, spilling 250,000 barrels of diesel fuel and killing thousands of krill and scores of penguins and other marine birds, and ruining several scientific projects along the coast. It was the worst environmental disaster to have hit the Antarctic. The next week, the British Antarctic Survey supply ship HMS *Endurance* hit an iceberg below its waterline near Deception Island and probably spilled a small amount of oil; on February 28, the Peruvian research vessel *BIC Humboldt* ran aground off Fildes Bay on King George Island and began leaking oil. Although the latter two events were relatively minor, they provided important support for the environmentalists' argument that the increased vessel traffic that would inevitably accompany mineral exploitation in the Antarctic would equally inevitably lead to increased risk of shipwrecks and oil spills.

All the while, in the background, environmental organizations continued to chip away at support for CRAMRA and to bolster the case for a mining ban.

The events of early 1989 helped their cause, but the knockout blow was a disaster at the other end of the world. On March 24, the oil tanker *Exxon Valdez* hit an iceberg and spilled at least 11 million gallons of oil into Alaska's Prince William Sound. It was a powerful, gut-wrenching illustration of the possible effects of a major oil spill in a polar region, and it proved to be the death knell for the minerals regime.

On May 22, Australia's prime minister, Bob Hawke, announced that his government would not sign CRAMRA. Instead, he said, Australia would seek to negotiate a comprehensive environmental protection convention and establish Antarctica as an international wilderness in which all mining and oil drilling would be prohibited. The following month, France also withdrew its support for CRAMRA. Given that all seven claimant nations needed to sign the treaty for it to come into effect, that double blow effectively killed it. In short order, other treaty nations—Belgium, China, Italy, India—expressed doubts about the minerals convention.

Domestic politics had certainly played a role in both the Australian and French decisions. France needed a good "green" policy issue to offset growing criticism not only over Dumont d'Urville but also of continued nuclear testing at Moruroa Atoll in the Pacific Ocean as well as lingering bad blood over French agents' sinking of the Greenpeace flagship *Rainbow Warrior* in 1985. It helped that recent parliamentary elections had placed some members of the Green Party into government and that the new environment minister had in earlier days sailed on a number of Greenpeace missions. Australia, too, had seen a surge of interest in "green" issues unrelated to Antarctic mining, including deforestation and damming in Tasmania and the threats posed to Australian beachgoers by a depleted ozone layer. In addition, subsequent analyses have argued that good old-fashioned concerns that CRAMRA diminished Australia's territorial claims may have influenced Canberra's decision.

But whatever the motivation, the emphasis now shifted from a regime that accepted the principle of mineral exploitation and sought to limit its adverse environmental effects to one that banned it altogether. On October 4, 1991, treaty nations signed the Protocol on Environmental Protection to the Antarctic Treaty, also known as the Madrid Protocol in honor of the city where it was finalized. Designating the continent a "natural reserve, devoted to peace and science," the protocol and its annexes contain strict regulations regarding such issues as environmental impact assessments, waste disposal, and wildlife conservation. A far cry from the days of cold war rivalry and jockeying for national position, it states:

> The protection of the Antarctic environment and dependent
> and associated ecosystems and the intrinsic value of Antarctica,
> including its wilderness and aesthetic values and its value as an
> area for the conduct of scientific research, in particular research
> essential to understanding the global environment, shall be fundamental considerations in the planning and conduct of all
> activities in the Antarctic Treaty area.

To this end, Article 7 of the protocol declares, "Any activity relating to mineral resources, other than scientific research, shall be prohibited." The ban is indefinite; the protocol cannot be opened to review or revision for fifty years after its entry into force.

Three years after the signing of the protocol, a huge piece of ice calved off the Astrolabe Glacier, near Dumont d'Urville. As it crashed into the water, it threw up a freak wave that swamped the islands of the Pointe Géologie Archipelago. The airstrip was severely damaged. It has not been rebuilt.

* * *

In 1992, considering its work there done, Greenpeace removed World Park Base. When I first set foot on Cape Evans a year later, only a plaque remained to remind the world that it had ever been there.

The environmentalists had won. Their nightmare scenario of oil exploration and development in the world's last great wilderness had, at least for now, been banished. Indeed, from being on the verge of allowing such drilling to take place, the nations of the world had completed an almost 180-degree turn and bestowed on Antarctica the most rigid and protective measures of any region on the planet. It was a famous victory.

At the other end of the Earth, it was—and remains—a completely different story. Far from receiving blanket protection from oil exploration, parts of the Arctic have seen some of the most extensive drilling outside of the Middle

East. It is not universally so: in the Canadian Beaufort Sea, for example, exploration and development were shelved because they were considered economically infeasible; in the Atlantic Arctic, such activities have thus far been largely limited to a couple of locations on the Norwegian shelf. But development is proceeding apace in parts of the Siberian Arctic, and oil production in Arctic Alaska is by far the largest industry in the forty-ninth state, now among the most significant oil-producing areas in the world.

The story of the battle over oil in the Alaskan Arctic stands in stark contrast to that in Antarctica. The political situation is, of course, very different: whereas Antarctica is part of the global commons, Alaska is one of fifty states in a sovereign country. Unlike the Antarctic, Alaska has a resident community, which largely lobbied in favor of oil development. And there never was any firm evidence of large oil fields in the Antarctic; in Arctic Alaska, however, there was plenty.

For environmentalists, however, the consequences of oil production in the Alaskan Arctic have in many ways mirrored the fate they predicted would befall the Antarctic were oil drilling ever to be allowed there. And not just in the Arctic itself. Even as some environmental campaigners celebrated their victory in Antarctica, others mourned the devastating effects of millions of tons of Arctic oil fouling one of the most stunning subarctic environments in the world.

Chapter 6

Crude Awakening

On a map it seems, at first glance, an insignificant place, a small dent in the coast of south-central Alaska. It is all but dwarfed by the gash torn to the west by Cook Inlet. But Prince William Sound is small only in relation to the enormity of the state in which it is located. Its surface-water area is around 2,500 square miles, roughly the same as that of Chesapeake Bay; the extent of the region as a whole, including the surrounding mountains, forests, and glaciers, is by some calculations about 15,000 square miles—an area greater than that of Connecticut, Delaware, Hawaii, Maryland, Massachusetts, New Hampshire, New Jersey, Rhode Island, or Vermont. Its countless bays and inlets yield more than 2,000 miles of shoreline, nearly equal to the coasts of California, Oregon, and Washington combined.

To its south, the sound opens into the Gulf of Alaska. On all other sides, it is hemmed in by mountains—the Kenai Mountains to the west, the Chugach Mountains to the north and east. Flowing down from the mountains and into the sound, twenty tidewater glaciers creak and groan, the inlets into which they enter, in the words of John Muir, "encumbered, some of them jammed and crowded, with bergs of every conceivable form, which, by the most active of glaciers, are given off at intervals of a few minutes with loud thundering roaring that may be heard five or six miles, proclaiming the restless work and motion of these mighty crystal rivers."

The climate of the sound is officially classified as maritime, although it might more prosaically be referred to as wet. There is an average of more than 100 inches of rain annually, and the fishing town of Cordova, in the sound's southeastern corner, receives more than 160 inches each year. Such an abundance of moisture allows the region to support temperate rain forests, but the area's ecosystems also include rocky beaches, lush meadows, and tidal flats. The sound's waters boast a plethora of fish and ten species of marine mammal, including whales, dolphins, porpoises, seals, sea lions, and more than 30,000 sea otters. There are thirty species of land mammal, from deer to black bears; about 3,000 bald eagles; and more than 200 species of seabird. Each year, in spring and fall, the Copper River delta hosts the largest seabird migration in North America as millions of swans, geese, terns, and other species congregate there on their way north or south.

The first people to reach Prince William Sound were Chugach Eskimos, but the dates and circumstances of their arrival are lost in the mists of time. Chugach legend recounts that the first encounter was with a sound much more heavily glaciated than today. Climatic records suggest this would have been several thousand years ago—and indeed, archaeologists have estimated the Chugach presence as dating back at least three or four millennia. The word *Chugach* may actually stem from a discovery by the Eskimos of an area of the sound that became ice-free following a glacier's retreat; or it may, like the names given to themselves by the inhabitants of the Arctic coasts, mean simply "the people."

Europeans did not enter the sound until the late 1780s, although several decades earlier, some did reach a point just beyond its entrance, pausing briefly before scurrying home. The members of that expedition were mostly Russian; their commander, Vitus Bering, was a Dane. Their arrival in the area marked something of a successful conclusion to what, for Bering, had been a convoluted and tiresome odyssey, one that began in 1725 when Peter the Great commissioned him to sail from the Kamchatka Peninsula "along the shore that runs to the north and that (since its limits are unknown) seems to be part of the American coast." Having then determined "where it joins with America," Bering was commanded to "sail to some settlement under European jurisdiction and, if a European ship should be met with, learn from her the name of the coast and take it down in writing, make a landing, obtain detailed information, draw a chart, and bring it here."

In 1728, three years after leaving St. Petersburg, Bering sailed from Kamchatka. A meticulous organizer, he was also an inherently cautious man—not necessarily the most desirable trait in an explorer—and as a direct result of such wariness, his first attempt was something of a failure. He sailed north through the strait that now bears his name, caught a glimpse through briefly parting fog of an island he called St. Diomede (now known as Big Diomede Island), and sailed as far north as the East Cape of Chukotka. Had he continued around and north of the cape, one part of his commission—the question of where (or whether) the two continents are conjoined—would have been answered there and then. But fearing his ship would be crushed by ice, he instead turned around and returned to Okhotsk and thence across land to St. Petersburg to report his findings. Such as they were: he had been gone five years and had little to show for his efforts other than a chart of the Kamchatka coast.

Widely rebuked, he requested another chance and was granted one; but partly at his own behest, this second search for America would begin only after Bering had survived a series of Herculean labors. Before he could leave port, he had first to travel overland from St. Petersburg and then, as Corey Ford wrote in his excellent narrative of the expedition, *Where the Sea Breaks Its Back,* to

> open up all of northeastern Asia, to introduce cattle-raising on
> the Pacific coast, to found schools for elementary and nautical
> instruction. He was to establish shipyards and ropewalks for the
> construction of his vessels, iron mines and foundries to turn out
> chains and anchors, lighthouses and storage depots and even a
> bimonthly postal communication service from St. Petersburg to
> Kamchatka and down to the border with China.

So exacting an undertaking was this that although Bering left St. Petersburg in 1733, it was not until 1741 that the expedition's ships were ready to set sail. It had been sixteen years since Bering first set out to satisfy Peter's curiosity, and by now he was weak, weary, and disenchanted. To add to the burden, barely had his ship left port than he fell gravely ill. His feet swelled up, his face lost its color; retiring to his cabin, Bering spent much of the journey staring without expression at the ceiling above his bunk, apparently having surrendered all interest in the voyage's outcome.

The expedition comprised two ships—the *St. Peter,* under Bering's command, and the *St. Paul,* captained by Aleksey Chirikov—and their first target was Gama Land, a large landmass said to lie southeast of Kamchatka. Bering was dubious and rightly so, for it does not exist; nonetheless, he had his orders. Several weeks were consumed in a fruitless search for the mythical land until Bering directed his crew to sail northeast toward what they presumed to be the unexplored terrain of northwestern North America. Finally, on July 16, 1741, the crew spied the shore of Kayak Island, southeast of Prince William Sound— their first view of Alaska. After all these many years, Bering had achieved what he had sought for so long; and yet, in a cruel irony, he was denied the glory of priority by other members of his own expedition. The two ships had become separated during the search for Gama Land, a venture Chirikov had also seen fit to abandon. The *St. Paul* had then headed northeast on a different track from the *St. Peter* and, either the day before Bering's ship came across Kayak Island or earlier that same morning, sighted what is believed to be Prince of Wales Island, in the Alexander Archipelago in southeastern Alaska.

It is doubtful, however, that had Bering known he had been trumped, his mood would much have changed. Even as his crew rejoiced in the achievement of finding what they had been looking for, their captain gave every impression of being a man for whom the discovery had come too late for him to care. Roused from his berth by the cheering on deck, he took one look at the island and ordered a boat to be sent ashore to secure fresh drinking water and then return. Only after a furious protest from Georg Wilhelm Steller, the ship's naturalist, did Bering consent to allow some scientific observations to be made; and so, leaping out of the boat in excitement as it hit the beach, Steller became the first white man ever to walk on Alaska soil.*

---

*Probably. To the southeast, Chirikov dispatched two boats from the *St. Paul* to investigate the new land, but neither was ever seen again. The most likely explanation is that they were caught in currents and destroyed, and their crews drowned; it is certainly possible, however, that they made it to shore.

As he stepped ashore, Steller saw foxes, ravens, and magpies and signs of fur seals and sea otters. He came across evidence of human habitation: discarded bones and the remains of a fire. The natives had presumably seen the white men arriving and fled, attempting to cover their tracks as they did so; even as he wandered around, Steller could feel unseen eyes watching his every move. An extraordinarily astute observer, Steller was able to deduce correctly, from the limited evidence before him, that the way of life and habits of the natives here were very similar to those on the Kamchatka Peninsula and that the two civilizations were therefore presumably related.

Steller pleaded for more time to continue with his discoveries; he had seen a column of smoke in the near distance and, given a few men and a couple of extra days, would surely be able to make contact with the natives responsible. But Bering's caution once more intervened; his instructions were to return to Kamchatka by mid-September, and he feared the weather would make the return journey longer than the outward. Without even allowing time for all the water casks to be filled, that very night he ordered that the anchor be weighed, and the *St. Peter* set sail for Kamchatka.

The voyage home was one that Bering would not complete. Increasingly ill, he transferred day-to-day command of the ship to his first officer, Lieutenant Sven Waxall; wracked now with scurvy, he lost much weight and all his teeth. On December 8, 1741, he died. He was buried on an island now known as Ostrov Beringa, one of the Komandorskiye Ostrova, or Commander Islands, just east of the Kamchatka Peninsula. By then, disaster had overwhelmed the expedition: scurvy raged through much of the crew, ultimately killing thirty-one of the ship's complement of seventy-six, and the *St. Peter*, blinded by fog and battered by storms, was wrecked on the shores of the island where Bering lies. After being forced to spend a wretched winter there, the survivors fashioned a crude forty-foot boat from the ship's wreckage and staggered back into Kamchatka in September 1742.

Their ordeal was over, but that of some of the wildlife they had encountered on the journey had barely begun. Wallowing in the waters around Ostrov Beringa had been enormous, blubbery creatures the like of which the Russians had never seen. They called the animals "sea cows." Today, we know these creatures were relatives of the manatees of Florida, the West Indies, West Africa, and the Amazon and the dugongs of the Indo-Pacific—placid herbivorous marine mammals that are themselves related to elephants. For the starving, desperate men of the *St. Peter,* however, the sea cows' taxonomy was of infinitely less import than the fact that they were endowed with copious layers of fat, and so the sailors waded into the water, laid waste to the creatures with hooks, gaffs, knives, and bayonets, and helped themselves to the blubber and flesh.

"We now," noted Steller, "found ourselves so abundantly supplied with food that we could continue the building of our new vessel without hindrance." The animals' blubber was "most delicious," and "beyond comparison with the fat of any marine animal, and even greatly preferable to the meat of any quadruped. . . . Melted, it tastes so sweet and delicious that we lost all desire for butter. In taste it comes pretty close to the oil of sweet almonds." As for the meat, it was "exceedingly savory and cannot be distinguished easily from beef."

And that, alas, would prove to be the sea cow's undoing. Apparently restricted in range to Ostrov Beringa and nearby Ostrov Mednyy, the sea cow was soon exterminated from both, and by 1768 it was extinct. Steller was the only scientist who ever saw it; no skins survive, and a half-dozen illustrations made by a member of the *St. Peter* party were lost and have never been found.

The people who hunted and ate the sea cow into extinction, however, were merely stocking up food to sustain them during expeditions in search of another quarry. Steller and his fellow survivors had returned to Russia with, among other things, the pelts of sea otters they had gathered during their enforced stay on Ostrov Beringa, and that cargo swiftly attracted the attention of the *promyshleniki,* a Siberian cadre of professional hunters and trappers who earned good money from the fur trade with China. The *promyshleniki* were ruthless, rapacious not only in their pursuit of furbearing animals but also in the ready plunder and subjugation of almost any peoples they encountered along the way, including the indigenous inhabitants of Kamchatka. They were in a sense true pirates, but of the land, not the sea; they were ill prepared to set out into the stormy, treacherous, poorly charted waters of the Bering Sea and North Pacific Ocean, but the luxurious fur of the sea otter proved temptation enough to lure them into the open ocean.

They built makeshift boats made from hand-hewn planks covered with walrus hide and known as *shitika*. Many were barely seaworthy, much like the hunters who crewed them; boat after boat became lost in fog or wrecked in storms, and yet, in wave after wave, the fearless *promyshleniki* swept onward.

They reached the uninhabited Komandorskiye Ostrova; from there, they attained the western tip of the Aleutian Islands. The first voyage to Ostrov Beringa returned with 1,600 otter pelts. An expedition to the Aleutian island of Adak procured 820; a visit by three ships to Tanaga Island produced 3,000; on Attu Island, 5,360 were secured by one vessel alone. The carnage was not limited to sea otters: fur seals and foxes were also relieved of their skins, sometimes in even greater numbers than the otters. The native inhabitants of the Aleutian Islands, too, felt the fury of the *promyshleniki,* and although many fought back in self-defense and in protest of the Russians' decimation of their favored resource, their actions frequently served only to prompt a disproportionate response from the invaders.

Island by island, island group by island group, the *promyshleniki* made their way, slaughtering sea otters and subduing and subjugating the Aleuts. Some of the Aleut men they killed; others they forced to hunt otters on their behalf while they made themselves at home with the native women. When the sea otters had been hunted to near-extinction in one place, the *promyshleniki* moved on, tracking steadily north and east toward what the Aleuts called Alyeska: the Great Land.

Over time, the random marauding of the *promyshleniki* gave way to more organized commerce, and in 1783, the crew of three ships under the command of Potap Zaikov, responding to rumors of a bay thick with sea otters, became the first recorded Russians to actually penetrate the waters of mainland Alaska. In many respects, Zaikov's expedition was a disaster: although the Russians had introduced themselves to the natives in the same way that had been tried and tested on the Aleuts—by raiding villages, stealing furs, and carrying off the women—the mainland Eskimos had proven a more determined foe and had attacked, killed, and wounded a number of the ships' crew. As a result, the Russians spent too much time fighting and not enough hunting and found themselves with too few pelts by the end of summer. They elected to spend the winter, but the Eskimos, not surprisingly, refused to reveal their hunting grounds or to provide any food, and as a consequence half the remaining men died of scurvy and starvation. However calamitous this initial voyage had been, Zaikov and his surviving crew were able to relate some promising news: the region was indeed, as Zaikov had been told, thick with sea otters. And so other Russian fur hunters followed in their path and headed to Prince William Sound.

* * *

Although the *St. Peter* had made landfall just outside the sound's entrance more than forty years previously, no Russians, so far as is known, had made it back prior to Zaikov. There is some evidence that maybe a few hardy *promyshleniki* had wandered that far north, in that the first expedition to definitively enter Prince William Sound encountered Chugach Eskimos who had iron spear tips and blue trading beads of presumably Russian origin. But it is just as probable that the Eskimos traded for them with other natives who had already encountered the Russians; therefore, the honor of being the first Europeans to enter Prince William Sound must properly go to the members of the expedition that first recorded having done so. And the commander of that expedition was none other than Captain James Cook.

After returning to England at the conclusion of his voyage to Antarctica, Cook had had little time to relax and rest on his laurels. For one thing, he almost immediately set to work writing his account of the voyage, which was published in May 1777 under the title *A Voyage Towards the South Pole, and Round the World, Performed in His Majesty's Ships the Resolution and Adventure, In the Years 1772, 1773, 1774, and 1775*. For another, the maps and charts of the world contained yet a good many blank spaces, and few if any explorers had proven as adept as Cook at solving the maritime mysteries of his time. Almost as soon as the *Resolution* and *Adventure* reached port, the Admiralty announced that the former would soon sail forth once more, this time to the North Pacific Ocean, to continue the quest for the elusive Northwest Passage. News of Bering's voyage had filtered through to England, as had a small book by Jacob von Stählin, secretary of the Academy of Sciences in Moscow, which included a map purporting to show the extent of discoveries made by Russian explorers and fur hunters in the North Pacific. Stählin's map included a semicircular string of islands, which he labeled the Aleutskians, but beyond that, much was based on equal parts speculation and imagination. The Great Land, Alyeska, of the Aleuts had been corrupted to a massive island, Alaschka, which Stählin had placed between the East Cape of Chukotka and a theoretical continuation of North America labeled Stachtan Nitada. Although the map was roundly ridiculed by the learned members of London's geographic circles, it was acknowledged that it contained a kernel of truth inasmuch as it was partly based on Bering's observations. There was at least a conceptual verity to Stählin's depiction of the Aleutians, and nobody had yet come up with more concrete evidence of what did lie north of the North Pacific. Perhaps Stählin was right; probably he was wrong. Either way, somebody needed to go and find out, and to the relief of the Admiralty, Cook volunteered for the task.

Cook dropped anchor in Prince William Sound on May 12, 1778. The visit was brief, functional, and unexpected, prompted by the mariner's need to find sheltered waters where the crew of the *Resolution* could fix a small leak the ship had sprung during a storm off Vancouver Island. Though unanticipated, it promised a great deal; both the *Resolution* and its consort, the *Discovery,* made contact with Chugach Eskimos—some of whom attempted to steal a couple of the Englishmen's boats and even boarded the *Discovery* before being chased away—and traded with them for sea otter pelts. For a while, Cook even considered that through this happenstance he might have actually stumbled across the Northwest Passage itself. But after sailing a short way to the north and dispatching two boats to explore farther north still, he determined that he was in an enclosed bay, and sailed on.

Cook often adopted the names native peoples used for particular geographic features and locations, but perhaps because the Eskimos he encountered did not volunteer one, he dubbed the area Sandwich Sound, after the Earl of Sandwich, first lord of the Admiralty. Sometime after the voyage, however, cartographers in London elected to rename it after Prince William (later King William IV)—perhaps a politically astute move at the time but not especially inspired, given that William was so little thought of even in his own country that he was frequently derided as "Silly Billy."

After leaving the sound, Cook rounded the Kenai Peninsula and explored the inlet, which he left unnamed but which the Earl of Sandwich, returning the favor, later designated Cook's River. (Not until a voyage led by George Vancouver in 1794 was the inlet's nature definitively established.) He came upon the Aleutian Islands and there encountered Aleuts and *promyshleniki*. He continued to the north, identifying the westernmost point of mainland North America and calling it Cape Prince of Wales, sailing through Bering Strait, and passing Norton Sound, Kotzebue Sound, and Point Hope, until finally the advancing pack ice blocked his path. This point of no return he dubbed Icy Cape, and then, after sailing west toward Siberia, he turned south. He had seen enough to confirm the total fallacy of Stählin's map—a map that "the most illiterate of. . . . Sea-faring men would have been ashamed to put his name to"— but had fallen short of confirming the existence of, let alone finding, the Northwest Passage.

It was a goal he would never achieve. He planned to winter in Hawaii (then known as the Sandwich Islands: being first lord of the Admiralty at the time of British naval domination and exploration carried considerable, albeit nominal, rewards) until March, head to Petropavlovsk on the Kamchatka coast, stay there until the end of May, and thence resume the search for a passage either to the northeast or northwest. But that plan, and the life of arguably England's greatest explorer ever, came to a sudden and shocking end at Hawaii's Kealakekua Bay: a conflict with natives, the discharge of muskets, a complement of marines overrun and fleeing, and suddenly, in the midst of it all, Cook struck in the back of the head with a club, wounded with a spear as he staggered forward, and then, overwhelmed, stabbed and clubbed repeatedly as he lay face down at the water's edge. On the evening of February 22, 1779, to the salute of cannon and firearms, his remains were laid to rest in the waters of the bay. The following morning, the *Resolution* and *Discovery* weighed anchor, leaving Hawaii, and their commander, behind.

✳ ✳ ✳

The ships sailed south. En route, they paused at Kamchatka—encountering Potap Zaikov and inspiring his voyage to Prince William Sound—and made a port of call at Macao, a Portuguese trading post on the Chinese coast. There, some of the town's Chinese merchants learned of the sea otter pelts on board the English vessels. When the sailors heard the exorbitant prices the Chinese were willing to pay for the furs, there followed a mad rush all over the *Discovery* and *Resolution* in search of any and all that could be found. So numerous had been the sea otters in Prince William Sound and Cook Inlet and so easy had been the fur to come by through trade with the natives that the ships were all but overflowing with pelts: pelts stuffed in lockers, pelts used as blankets and mattresses. On board, they had become curiosities, trinkets, attractive souvenirs of an epic voyage, but little more. Now, suddenly, they were valuables, with sailors selling their stocks for hundreds of dollars—a fortune at the time for a crewman of the lower decks. Even those pelts that for months had warmed and comforted their owners in their bunks fetched top coin. So excessive was the money to be made that "the rage with which our seamen were possessed to return to Cook's River, and, by another cargo of skins, make their fortune at one time, was not short of mutiny."

Almost as much as the news of Captain Cook's death, and certainly more so than his inconclusive efforts to determine the existence of a Northwest Passage, the big story in England by the time the *Resolution* and *Discovery* reached home was the copiousness of sea otters in Alaska's waters and the extraordinary wealth to be garnered from the sale of their pelts. The Russians, concerned about the potential for international rivalry, established permanent settlements in Alaska to give themselves a competitive advantage and to lay down a marker by which they could make formal claim to what they called Russian America. Ultimately, Britain and the United States signed treaties recognizing Russian sovereignty and mostly contented themselves with the otter-rich islands, bays, and inlets of southeastern Alaska, leaving control of Prince William Sound largely to the Russians.

Within a matter of decades, however, the Russians were gone. Sea otter populations had been massively depleted, and the rapidly diminishing returns meant that Alaska was a luxury that Russia, its resources stretched by bloody conflict with England over Crimea, could no longer afford. In 1867, Russia sold Alaska to the United States for roughly two cents per acre.

* * *

Before the arrival of Europeans, Alaska's waters contained perhaps 150,000 sea otters. By the time the species was granted full protection from hunting in 1911, it was almost too late. The otter had been extirpated from much of its range, and only a few isolated populations—maybe 1,500 animals in all—

clung on. But those few populations in the region began to grow and expand their range again, and over time the species made a remarkable recovery. Although estimates were hard to come by, it was reckoned that by the mid-1980s the sea otter population in the state certainly numbered in the tens of thousands and may even have been close to its pre-exploitation level.

Prince William Sound alone boasted perhaps 35,000 otters; it was scarcely possible to look anywhere in the sound's waters without seeing them, lying on their backs or quizzically poking their whiskered heads above the surface, and even harder to imagine that they had once been so close to extinction. But the story was not over just yet.

"It's Another Kuwait"

There is just one way in and out of Prince William Sound by road, and that is via the town of Valdez, in the sound's northeastern corner. Starting from the town's center, the Richardson Highway heads roughly north until, about 115 miles out, it meets the Glenn Highway. The Glenn leads to nearby Glennallen and on to Anchorage; the Richardson continues north to Fairbanks. From Fairbanks, its roughly northward trajectory is maintained first by the Elliott Highway and then by a stretch of road officially known as the James Dalton Highway but widely known in Alaska as the Haul Road.

The Haul Road can be rough going; if you travel it, you would be well advised to carry a couple of spare tires. Stock up on supplies, too, and take some extra gasoline, especially in winter: there are few services along the way, and none at all for the final 225 miles. But if you stick it out to the end, you will find—about 800 miles north of Valdez, well above the Arctic Circle, near the coast of the Beaufort Sea, a couple hundred miles west of the Canadian border—a town of sorts called Deadhorse.

There really is not a great deal to Deadhorse, or a great deal to do there—a few motels, a couple of restaurants, a store, and a post office. There is no legal liquor. In the Prudhoe Bay Hotel, there are a number of tanning beds, a Ping-Pong table, a pool table, and a few easy chairs arranged in front of a large-screen television set. The store sells an eclectic combination of hardware, magazines, videotapes, souvenirs, clothing, and a dazzling array of pornographic toys, including blow-up dolls of several animal species and of human dwarfs.

The door to the store opens outward, and posted on the inside is a handwritten note warning customers to take a good look before stepping out into the street: polar and brown bears are frequent visitors here, and as the sign says, "although they look cute," they are also big and fierce and come equipped with sharp teeth and claws.

Deadhorse is no metropolis, but it was not designed to be one, or even a place to grow any kind of roots. Its nature becomes clear after a cursory glance at the planes taking off and landing at the airstrip. A few commercial flights come through here—Alaska Airlines, mostly—but just as many are charters: planes decked, for example, in green and yellow livery and sporting the logo of BP, or the British Petroleum Company (now BP Amoco). Deadhorse is an oil town.

✳ ✳ ✳

At least as early as the nineteenth century, the Iñupiat discovered oil seeps near Cape Simpson, fifty miles southeast of Barrow, and at Angun Point, about thirty miles southeast of the village of Kaktovik, near the Beaufort Sea coast; the area's inhabitants frequently traveled to the seeps and cut blocks of oil-soaked tundra for use as fuel. In 1886, an exploratory expedition by the U.S. Navy found oil near the upper Colville River, about two-thirds of the way between Barrow and Deadhorse; a little less than thirty years later, a teacher in the Iñupiat village of Wainwright discovered two springs of oil just shy of the Arctic Ocean coast near a place called Smith Bay. Five years after that, in a report on the Canning River region, about sixty miles southeast of Prudhoe Bay, geologist Ernest de K. Leffingwell noted the existence of oil seeps there; and in 1922, geologist Alfred Brooks was quoted as predicting that "oil will be found in Alaska, and the probabilities are that there are extensive [oil] areas in the Territory."

The first commercial strike in Alaska actually came much farther south, at Swanson River on the Kenai Peninsula in 1957, but oil companies continued to suspect that the largest fields were most likely in the Arctic—a region that, because it slopes downhill from the mountains of the Brooks Range to the shores of the Beaufort Sea, is known as the North Slope. During World War II, the federal government had withdrawn most of the lands north of the Brooks

Range from public or private ownership and use, but after Alaska acceded to the Union in 1959, it was granted a total of 103,350,000 acres to select as state lands—about 28 percent of the total land area and an extent almost twice as large as North and South Carolina combined. And acting on the advice of geologists who assured him it would provide the new state with the greatest opportunity to reap a financial windfall, the first land that Alaska's inaugural governor, William Egan, selected was the North Slope.

Egan had been advised that potentially the greatest oil fields lay in the northeastern quadrant, but that had already been designated as the Arctic National Wildlife Range, and the federal government would not release it. He could not take the area to the northwest either because that was locked up as a petroleum reserve. But he could and did select the area in between, and in December 1964 the state of Alaska held its first competitive lease sale, for 600,000 acres in the Colville River area. The bulk of the leases were snapped up by BP and its partner Sinclair Oil Corporation, but the real prize was yet to come: the lease sale, scheduled for the following July, centered on Prudhoe Bay.

Prior to the lease sales, teams of geologists from rival oil companies descended on the North Slope, fighting fierce Arctic winter as they fanned out across the frozen tundra in search of signs that the ground was sheltering large fields of oil. The oil companies prided themselves in operating in some of the harshest environments, but the Arctic presented challenges much different from, and far greater than, anything they had encountered in the sands of Arabia or the jungles of Venezuela. Their work was complicated by the fact that the state prohibited them from working during summer for fear that heavy machinery would permanently damage the tundra. As a result, survey teams were obliged to labor through the long polar night, with temperatures regularly plummeting to −40°F and fearsome wind chills making it seem, if such a thing were possible, still forty degrees colder. Beneath the surface, the drills had to chew their way through the permafrost—the layer of permanently frozen soil that is almost a defining feature of the Arctic and that around Prudhoe Bay is frequently more than 2,000 feet thick.

But operating equipment in an Arctic environment was only half the problem. First, the companies had to find a way to get it there. The Beaufort Sea is

open to shipping for only three months of the year—and that, naturally, during summer, when the drillers could not operate. Most companies used airlifts
to fly dismantled equipment to the frigid north, where it had to be reassem-
bled amid howling winds and whiteouts. BP employed a Canadian drilling
team, which barged its heavy equipment from the Mackenzie River to the
Beaufort Sea and then up the Colville River to a site forty miles inland. And,
daringly, Richfield Oil Corporation—the same small company that had struck
oil at Swanson River—launched an ambitious modern-day wagon train: a caravan of Caterpillar bulldozers hauling bunkhouses, drilling apparatus, food, and
supplies.

It was a hazardous, trying journey: almost 500 miles from Fairbanks to a
staging area at Sagwon, across little-known terrain and without the advantage
of anything even remotely resembling a road. The first sixty miles or so of
travel took eighteen days, during which one of the Cats broke through the ice
and settled into six feet of water and mud in a previously unseen beaver pond.
Attempts to winch it out resulted in snapped cables and frayed tempers; only
by dynamiting the entire pond was the crew able to free the entombed vehicle.
But passions cooled, and the remainder of the journey was completed in just
twenty-two days—four of which were spent hunkered down waiting out a furious blizzard. Finally, the caravan, the very first expedition ever to bludgeon its
way overland to the North Slope, arrived at Sagwon, and there it met up with
an excited team of Richfield geologists.

The geologists were convinced they had found a major reservoir of oil in
the Prudhoe Bay area; so much so, in fact, that although most oil companies
were obsessively secretive about the results of their surveys for fear of giving
their rivals a competitive advantage, Richfield actively sought a partner out of
concern that it alone would not be large enough to take advantage of the vast
oil deposits it anticipated. As a result, the company hooked up with Humble
Oil and Refining Company (which would later become the Exxon Corporation), and on July 14, 1965, the two bid as high as $94 per acre—more than
twice BP's average bid—and came away with roughly two-thirds of all the
Prudhoe leases.

Early in 1966, Richfield merged with the Atlantic Refining Company to
become the Atlantic Richfield Company, or ARCO; and beginning in February

of that year, the combined company began digging its first well, nicknamed Susie. Amid much excitement and anticipation, the rig began to drill into the Arctic ground: at 1,800 feet, it caught its first suggestion of oil, but that proved to be only a hint and nothing more. At 3,500 feet there was another trace, but again it proved a false alarm. The drill continued downward: to 10,000 feet (almost two miles) and on to 13,500 feet, until finally it was plugged. It had found nothing. Susie was dry.

By this time, despite the early euphoria, Susie was the only major rig remaining on the North Slope. BP and Sinclair had already struck out at the wells they had drilled at Colville, a disappointment that was enough for Sinclair to pull out of the North Slope altogether. BP contemplated doing likewise but had decided to sit on its leases and wait to see what Susie brought forth. When it found nothing, ARCO, too, stood poised to go home. But a few persistent voices in the company argued for one more chance. ARCO would have to pay to dismantle and transport the rig anyway, so before going to that expense, why not have the geologists recommend what they believed to be the most promising spot, use the Cats—which had required so much effort to reach the North Slope—to move the rig to that spot, and take one final shot? The company's board agreed; but, as chairman Robert O. Anderson later recalled, if "the Prudhoe well had been dry, we were going home."

It was not, and they did not. Drilling began in earnest after the winter freeze-up, in November 1967; the day after Christmas, a burst of gas shot out the end of a flow pipe, ignited, and erupted in a fifty-foot flame. For the best part of two days it burned, lighting the Arctic night for miles around and venting with a deafening roar. Where there is gas, there is not necessarily oil, but where there is that much gas, the odds of oil are high. On February 15, 1968, ARCO went public. "It looks extremely good," said Harry Jamison, the company's general manager for Alaska. "We have a major discovery." By March, ARCO was more specific: although further tests had been "inconclusive," the well was now producing 1,152 barrels of oil per day.

ARCO moved the rig seven miles southeast to the Sagavanirktok, or Sag, River and drilled a second, confirmation well. The crew found what they were looking for, and now there was no doubt: ARCTIC FIND IS HUGE, ululated the *Anchorage Daily Times,* and so it was. By now, the first well was producing 2,415 barrels per day, and the Sag River find was doing even better, with 3,567 barrels daily. Subsequent wells yielded 20,000, even 30,000, barrels per day—on a par with some of the most productive fields in the Middle East. Noted *Business Week:* "'It's another Kuwait,' say the optimists in the oil industry, and they just may be right."They were not far wrong: at roughly forty-five miles by eighteen miles and with estimated total reserves of 20 billion barrels, it was by far the largest oil field ever found in North America, and one of the largest in the world.

* * *

In November 1968, a team of bulldozers rumbled out of Livengood, about 60 miles north of Fairbanks. Over the course of almost four months, roughly following the trail blazed by Richfield's Cats several years earlier, they thundered slowly toward the North Slope, cutting a 400-mile road through the tundra, across the Brooks Range and all the way to Sagwon.

The road was the brainchild of Walter J. Hickel, an Anchorage developer who, in a stunning upset, had been elected Alaska's governor in 1966, besting incumbent William Egan by a margin of just more than a thousand votes. A man of undeniable vision, Hickel had, during his election campaign, lambasted the first generation of Alaska's political leaders for their "timidity" and offered that Egan was "a little bit afraid of bigness."That was a crime of which no jury could find Hickel guilty, and immediately upon election, he set about keeping his promise to "get Alaska moving." He urged the "opening up" of the Arctic and sought to "forever banish the myth that the Alaskan Arctic is a land of ice and snow." It is not, he insisted, "the harsh kind of region people think it is." He spoke often and admiringly—as he does today—of the Trans-Siberian Railroad and its success in opening up Russia's remote Arctic reaches, and he established the Commission for Northern Operations of Rail Transportation and Highways (NORTH Commission) to explore the feasibility of extending the Alaska Railroad to Nome and Kobuk. The commission's mandate also included consideration of construction of a winter road to the North Slope, but with the discovery of oil, the governor decided that time was too precious to wait on the findings of some committee. Reasoning that increased accessibility could only accelerate Arctic Alaska's development, Hickel gave the go-ahead for the bulldozers to hew a path toward Prudhoe.

There is a way to construct a winter road in the Arctic: build up snow and cover it with water to form ice. But Hickel did not want a winter road to the slope; he wanted a year-round route. And so his bulldozers, instead of building up, dug down and carved a trench through the tundra. With the protective blanket above it ripped away, the permafrost was exposed to the returning summer sun and rapidly started to melt; puddles grew to ponds, ponds expanded into veritable lakes, and the entire thing turned into one giant, unusable ditch stretching halfway across Alaska.

By the time construction was under way, however, Hickel had been transferred to Washington, D.C., having been tapped by President Richard Nixon to be his new secretary of the interior. His successor as Alaska's governor, Keith Miller, anointed his predecessor's bequest the "Hickel Highway." "This impossible road," he announced, "shall be known by the name of the man whose courage, foresight, and faith in the Great Land gave Alaska what surely will be one of its greatest assets."

He might well have referred to it as impassable rather than impossible: the road was in use for only six weeks, during which time it transported just seven and a half tons of heavy equipment, which could just as cheaply—and much more easily—have been flown to Sagwon. Far from one of Alaska's greatest assets, it was one of the state's whitest elephants. The "Hickel Canal," critics derisively dubbed it, but behind their mockery there was genuine anger.

The scars etched in the tundra by Hickel's impulsiveness were permanent, and although Hickel was unrepentant—"So they've scarred the tundra. That's one road, twelve feet wide, in an area as big as the state of California"—his lack of contrition only elicited more outrage.

Such outrage was not prompted solely by the immediate effect of the road itself, although that was enough. Nor was it caused by the lack of public notice, comment, and participation—although that, too, fanned the furor. As much as anything, it was a response to what the road represented.

This was not the first instance of conflict surrounding Alaska's nascent oil industry; development of Richfield's strike on the Kenai had been delayed by legal challenges from environmentalists on the grounds that the oil had been found in a region originally designated as the Kenai National Moose Range.

But the building of the road, and the sudden, mad scramble to develop what was in the minds of many the last true expanse of wilderness remaining in the United States, represented, in the words of an article in *Field and Stream* magazine, an "ugly symptom of the fever that has Alaska by the throat." One observer called it "the first violent change, the first major intrusion of the modern industrial world" into one of the few remaining regions where that world had never left its mark. It was, noted one commentator, the "end of thousands of years of solitude."

From the Arctic to Prince William Sound

In the darkness of midnight on June 16, 1903, a thirty-two-year-old, forty-seven-ton sloop slipped out of Christiania Harbor in Norway, down a fjord, and into the open sea. The ship, the *Gjøa,* was named after the wife of the former owner; when a new owner bought the ship in 1900, he elected to retain its appellation. That new owner was Roald Amundsen, and the *Gjøa* was his first command.

The ship sailed west, crossing the northern Atlantic Ocean and entering the myriad channels of the Canadian Archipelago. On September 9, Amundsen spotted a landlocked cove on the southern coast of King William Island in present-day Nunavut and anchored a short way offshore. On October 1, ice began to form around the ship, and by October 3, the *Gjøa* was frozen in. For two winters, the ship and its small crew stayed there, interacting with and learning from the Netsilik Eskimos who came to greet them. Finally, on August 13, 1905, Amundsen sailed out into Simpson Strait and then west through the uncharted waters of the Canadian High Arctic. Less than two weeks later, on the morning of August 26, the *Gjøa* encountered a whaler, the *Charles Hanson* sailing out of San Francisco, approaching the Norwegian vessel from the west. When Lieutenant Helmer Hanssen—who a little more than six years later would stand with Amundsen at the South Pole—burst into his cabin early in the morning, Amundsen was initially annoyed at being woken so soon after retiring to his bunk following his night watch. But when Hanssen yelled, "Vessel in sight, Sir," the captain immediately understood the words' significance. He had been the first to achieve the goal sought by so many for so long. He had sailed through the Arctic from east to west.* Three and a half centuries of searching were over. The *Gjøa* had sailed the Northwest Passage.

It would be thirty-five years until the next successful crossing: the Royal Canadian Mounted Police schooner the *St. Roch* traveled the route east between 1940 and 1942 and in the opposite direction in 1944. Thereafter, there were other firsts: HMCS *Labrador,* the third vessel to traverse the passage, was the first warship to do so, completing the journey as part of the first continuous circumnavigation of North America; the USCGC *Storis,* accompanied by the buoy tenders *Bramble* and *Spar,* composed the first squadron, and were the first U.S.-flagged ships, to make the trip; the nuclear submarine USS *Seadragon,* cheating somewhat, steamed under the ice in 1960. But there had still been only a handful of successful traverses—ten, in fact, by nine vessels, including two submarines—when in August 1969 the SS *Manhattan* eased out of port in Chester, Pennsylvania. At roughly 150,000 tons and powered by a 43,000-horsepower engine, the *Manhattan* was a behemoth, the largest vessel in the U.S. merchant fleet, converted at the expense of Humble Oil into a would-be icebreaker. Its hull had been reinforced with steel plating and its bow widened so it could break a wide path through the ice.

Humble had gone to this trouble and expense for one simple, and important, reason: to determine the feasibility of using icebreakers to transport North Slope crude through the Northwest Passage. Although Humble and Richfield had combined to buy leases in Prudhoe Bay, it had been a temporary marriage of convenience; once the oil was out of the ground, it was everyone for himself as each oil company sought a competitive advantage in terms of distribution and sales. BP and ARCO favored the idea of delivering the oil from Prudhoe via a pipeline that would snake south across Alaska to Prince William Sound, from where it could be transported by tankers to the West Coast of the United States. As well they would: the West Coast was where both companies

---

*Well, almost. The final stretch would have to wait a little while longer because ice barred his progress and forced him to delay for one more winter, just east of Herschel Island. It would be another year before the *Gjøa* rounded Point Barrow, but to all intents and purposes, the meeting with the *Charles Hanson* was the sign that the Northwest Passage had been achieved.

were dominant, and such a route would enable them to control much of Alaska's oil distribution. Humble, on the other hand, ruled the roost in the east, and it saw the Northwest Passage as perhaps its best and only chance to interrupt the other companies' hegemony.

The experiment, however, was not a great success. Amundsen had deliberately selected a small vessel for his assault on the passage, reasoning, in the words of his biographer, Roland Huntford, as follows: "He faced narrow waters and treacherous shoals; safety lay in a nimble ship with shallow draft. Others had come to grief trying to crush their way through the ice; he would push his way between the floes." This did not augur well for the *Manhattan,* by far the largest ship ever to attempt the passage; and sure enough, despite the advantage of two helicopters flying ahead of it to scope out a suitable route, the *Manhattan* frequently encountered heavy ice, from which it had to be extricated by the two icebreakers—one U.S., one Canadian—accompanying it. Freed it was, however, and able to continue its journey until, on September 19, it reached Prudhoe Bay.

On the way back east, the *Manhattan,* carrying a single, symbolic barrel of North Slope crude, was struck by ice, which tore open a gash "big enough to drive a truck through" in the hull and into one of the tanks assigned to carry oil. But by then, it was already clear that the bays of the Beaufort Sea were too shallow for tanker traffic and the building of a tanker terminal in that environment would be impossible. By 1970, Humble had decided the whole idea was impractical.

Many others applied themselves to the question of how to deliver crude oil from the Arctic to the refineries and gas stations of the outside world. The Boeing Company proposed a fleet of giant aircraft, with wingspans in excess of 478 feet, powered by twelve engines of the kind that keep 747s aloft. The General Dynamics Corporation announced plans to refit nuclear submarines as subsea tankers. Others nominated a railroad extension, a pipeline through Canada, or, noted Daniel Yergin in *The Prize,* his Pulitzer Prize–winning history of the oil industry, "a fleet of trucks in permanent circulation on an eight-lane highway across Alaska (until it was calculated that it would require most of the trucks in America)."

But from the earliest days following the strike at Prudhoe, the three main oil companies—ARCO, Humble, and BP—had formed a loose consortium, the name of which—the Trans-Alaska Pipeline System, or TAPS—reflected their preferred method of transport. The alliance was not always easy—as indicated, for example, by Humble's flirtation with the Northwest Passage. But for BP and ARCO especially—and, with the failure of the *Manhattan,* also for Humble—a pipeline was the cheapest, easiest, and most practical way to get oil from the North Slope to the rest of the world.

The plan the companies hatched, and formally announced on February 10, 1969, was a bold one indeed. The pipe would start at Pump Station No. 1, near the middle of the Prudhoe Bay oil field, and then stretch some 800 miles south, crossing high over the mountains of the Brooks Range and all the way to the port of Valdez, the northernmost ice-free port in the United States. George W. Rogers of the University of Alaska described some of the extraordinary natural obstacles the pipeline's construction would have to overcome:

> The line will reach altitudes of 4,700 feet through the mountain passes of the Brooks Range and of 3,500 feet through the Alaska Range; approximately three-fourths of the distance will be over permafrost . . . ; and the last fifty miles through the Keystone Canyon is described as "some of the most difficult mountain pipeline construction that has ever been undertaken." The many rivers to be crossed are a roll call of giants, the largest being the Yukon, which has never been spanned. Most are subject to heavy flooding and dramatic ice break-ups in spring.

The state had already selected land at Valdez for an oil pipeline terminal, and the Forest Service granted TAPS additional land within Chugach National Forest. And the TAPS consortium (which was later expanded to include all seven oil companies with interests on the North Slope and formally incorporated as the Alyeska Pipeline Service Company) had selected the route with little or no input from outsiders. It had approached the necessary officials and government departments only when seeking requisite rubber-stamping, and within two months of the formal announcement, it had bought 800 miles of forty-eight-inch-diameter pipe from three Japanese companies. Everything, it seemed, was progressing smoothly; but if the oilmen reckoned it was all a done deal, they reckoned too soon.

Already, noted many critics, the tundra of the North Slope had been damaged by the tracks of Cats and the activities of oil drilling. Writing of a series of visits to the area, Edgar Wayburn, vice president of the Sierra Club, recorded: "I saw thousands of oil cans and almost no wildlife. I saw great vistas of scarred tundra: cat tracks, roads bulldozed out of the tundra, seismic lines in crazy patterns everywhere, and dozens of air strips to accommodate over 1,120 air operations a day. And this, as we know, is only the beginning."

Some of the oil companies and many of their associates portrayed themselves as sympathetic to environmental concerns. "In North America, the petroleum industry has, by its own initiative, and by government regulation, set a high level of standards for field operations," insisted Russell A. Hemstock of Imperial Oil Limited. "These standards will be maintained in the Arctic and know-how will be developed to operate safely in this new and severe environment." At the same time, however, some of the pipeline's fiercest and most high-profile advocates mocked the notion that the tundra was even worth protecting. In his outstanding review of the debate, *The Trans-Alaska Pipeline Controversy: Technology, Conservation, and the Frontier,* Peter A. Coates recalled, for example, the correspondent for the *Christian Science Monitor* who insisted that the pipeline would "run through a bleak, barren, hostile, unpopulated country. If there is anyplace in America where a pipeline would not damage anyone or anything, this is the place," and the editor of *Alaska* magazine who decried the North Slope as "the most desolate area in the world. If I never see it again, it's fine with me."

Sometimes the pipeline's proponents simply invited criticism. The plan, as originally announced, was for the entire length to be buried—a specification that, if meant to demonstrate a desire to avoid damaging the Arctic environment, served only to illustrate a profound ignorance about how to go about doing so. As had Wally Hickel and his bulldozers, the TAPS consortium had reckoned without permafrost, and specifically the consequences thereon of placing a pipe in the middle of it containing oil at a temperature of 170°F. The result, several experts noted, would be substantial melting of permafrost, as a result of which the pipe would "float" to the surface: one predicted that "we could expect to see buried pipe swinging in the air."

As a result, talk turned to raising large sections of the pipeline aboveground, but that approach also came under assault. It could result, noted George W. Rogers, in "an ecological Berlin Wall about six feet high dissecting the mainland mass of the state, accompanied by the constant threat of an inland version of [the 1969 oil spill off] Santa Barbara if leaks developed." Of particular concern was the possible effect of such a barrier on caribou migration patterns. The U.S. Fish & Wildlife Service noted:

> In the Arctic region, caribou normally spend the winters in the lower mountain areas and then in May move to the North Slope, where they bear calves. Later they move into the mountains for the remainder of the summer. The subarctic herds seasonally migrate between the river valley and high mountain areas. As is obvious, a forty-eight inch pipeline placed on the flat terrain of the North Slope would form a fence or barrier where it crossed migration routes.

For supporters of the oil pipeline, the constant rain of criticism was infuriating, distracting, and unnecessary. "God placed these things beneath the surface for a purpose," groused Mayor Julian Rice of Fairbanks. "For us to say that we shouldn't use them is to be anti-God." The rage of Alaska senator Ted Stevens was more secular: "I am up to here with people who tell us how to develop our country." Proponents waved away concerns about the environmental effects of pipeline construction, stressing that the pipeline corridor would occupy just sixty square miles, or just 0.01 percent of the total area of Alaska. In the words of Vide Bartlett, widow of Alaska senator Bob Bartlett: "If you took a daily newspaper, laid it out to one full double page and stretched across it a piece of ordinary black thread, you'd have some idea of just how much the pipeline is going to take."

To warnings such as that of David Brower of Friends of the Earth—"If oil is pumped out of Prudhoe Bay and then shipped down the west coast, we will eventually have an oil spill leading to the greatest kill of living things in history"—the pro-development *Anchorage Times* editorialized, "The fears about damage from oil spills are like the fears of Henny Penny when she ran to tell the king that the sky was falling." Ted Stevens reportedly assured concerned Cordova fishermen that "there will be not one drop of oil in Prince William Sound."

And when environmentalists argued that the fundamental changes almost certain to result from the growing infrastructure and the establishment of settlements and associated roads and airstrips would change Alaska forever, proponents retorted that that was precisely the point. Vide Bartlett testified before Congress that "until the discovery of oil in 1957, we were still begging for our bread." She continued:

> We have always known that our hope for self-reliance lay in development of our natural resources. We must not be locked in a cavern of ice without the benefit of warmth from the use of our very own resources. . . .
>
> This is not all oil versus wilderness. There are real live people in this equation, with children, futures, and hopes. Do not forget them! Let's get on with the job.

Not every Alaskan shared that viewpoint, however. And barely had Bartlett's words been uttered than Congress heard a counterpoint, an affirmation that it was not only environmentalists who were hostile to the oil companies' plans. Alaska Natives had, once more, been largely ignored in the rush to develop the region's mineral riches. The federal Bureau of Indian Affairs (BIA), for example, had leased land belonging to the village of Tyonek, on the coast of Cook Inlet, to the Pan American Oil Company without informing the village and had placed the money in a government fund, again without the villagers' knowledge or consent. When the state of Alaska submitted its claim for the North Slope, it asserted that the area was free of aboriginal use and occupancy, notwithstanding several thousand years of residency by the Iñupiat and their predecessors. And Senator Ted Stevens similarly dismissed concerns over oil development in the Arctic by proclaiming that "there are no living organisms on the North Slope." Many natives vowed, however, to make their voices heard. One such voice, that of Charlie "Etok" Edwardsen, an Iñupiaq from Barrow, resounded powerfully after Bartlett and other pipeline advocates had finished speaking. "How can white men sell our land when they don't own it?" he exclaimed. "How can they lease our land to oil companies when they don't own it? If the pigs want to use our land, let the pigs pay the rent."

Beginning in 1961, native groups had begun filing their own land claims, and by 1966 they had claimed some 370 million acres—because of overlaps, a greater amount than the actual total land area of Alaska. The barrage of claims brought the system of state and federal leases to a grinding halt and prompted Secretary of the Interior Stewart Udall to impose a land freeze: there would be no more oil leases until the native claims issue was resolved.

The pipeline's fiercest advocates reacted angrily. George Moerlein of the Alaska Miners Association asserted that "neither the United States, the State of Alaska, nor any of us here gathered as individuals owes the Natives 1 acre of ground or 1 cent of the taxpayers' money." In a radio address, Governor Walter Hickel protested that "just because somebody's grandfather chased a moose across the land doesn't mean he owns it."

Disregarding such insults, on March 9, 1970, five native villages claimed the ground over which the Trans-Alaska Pipeline and its supply road would pass, and asked the federal district court to block issuance of a permit for the road to be built. A few days later, three conservation groups sued the Department of the Interior, asking for the TAPS project to be halted on the grounds that, among other things, it violated the newly adopted National Environmental Protection Act (NEPA). On April 13, 1970, federal judge George L. Hart Jr. filed a temporary injunction against the TAPS project.

The pipeline now was in genuine peril, enough so to persuade the oil companies it was time for them to lobby for a resolution to the claims issue. On December 18, 1971, after a year and a half of negotiations, compromises, and drafts, President Nixon signed into law the Alaska Native Claims Settlement Act, or ANCSA. ANCSA extinguished all native claims based on aboriginal title in Alaska; in return, it gave the state's estimated 75,000 natives $962 million in cash and 44 million acres of land. Forty-five percent of the reparation money would be administered by villages and the rest by thirteen regional corporations, including one "at-large" corporation for nonresident natives. In addition, villages would own the surface rights to their lands and regional corporations would own the subsurface or mineral rights.

To some native activists, such as Charles Edwardsen, ANCSA was a sellout; to critics, it surrendered far too much. Either way, the native claims issue was now resolved. The path still was not clear for construction of the pipeline, however. In the suit they filed that brought TAPS screeching to a halt, environmentalists charged, and the courts agreed, that no progress on the pipeline could be permitted without a comprehensive environmental impact assessment. Following Judge Hart's ruling, the Department of the Interior set to work on producing just such an assessment, and it released a draft in January 1971.

As is often the case with such documents, there was much to disappoint all sides. Pipeline advocates were irritated by the draft's acknowledgment of "the fundamental change that development would bring" to the North Slope and that "the original character of this corridor area in northern Alaska would be lost forever." The report also accepted that there would inevitably be some oil spills. At the same time, however, it argued that the ecological damage would be minimal, that caribou would prove "adaptable," and that a pipeline across Alaska would be the least environmentally damaging and most economically viable way of transporting oil to the rest of the United States, particularly to the West Coast.

The draft assessment was only the start of a process. Its publication was followed by a series of public meetings in Anchorage and Washington, D.C., the collection of more than 12,000 pages of testimony in thirty-seven volumes of commentary, and the release of a final statement in March 1972. Although this version added extra emphasis to some of the ecological damage that was likely to occur if TAPS were built—enough for Ted Stevens to complain that "it was not written by a proponent"—its fundamental position remained the same: the pipeline should be built. On August 15, 1972, Judge Hart dissolved his injunction, declaring that the environmental impact statement process met all of the Interior Department's requirements under NEPA. The Nixon administration took an increasingly activist role in advocating the pipeline's construction. And then the oil companies started turning the screws: the United States, they insisted, was in the early throes of an "energy crisis," and if the pipeline were not built, the country would face catastrophe.

Although some questioned the reality of this "crisis"—an article in the journal *Foreign Policy,* for example, argued that it was a fiction generated by the oil companies and oil-exporting countries for their own benefit—the prospect of massive energy shortages was enough to turn the tide of public opinion outside Alaska, which hitherto had largely been anti-TAPS, strongly in the other direction. Democratic senator Frank E. Moss of Utah declared, "I cannot get overly upset about observing the ritual of the mating season for Alaskan caribou when in the city of Denver last weekend it was almost impossible to find gas," a statement echoed by California congressman Craig Hosmer: "To preserve the 7,680 acres that would be occupied by the pipeline seems to me an inordinate price to pay for fuel rationing, cold homes, cold schools, and blackmail by the Arab world." Had the pipeline been built in 1969 as originally planned, asserted Hosmer, the wave of shortages and brownouts that occurred over the summer of 1970 would never have materialized. Plenty of voices, including those not inherently hostile to TAPS, tried to argue that Alaska oil would not in itself make much of a dent in any perceived energy shortage, but they were finally drowned out by an event that occurred thousands of miles away.

On October 6, 1973—Yom Kippur, the holiest of Jewish holidays—Egypt and Syria launched aerial attacks on Israel, and, nominally in retaliation for U.S. support of the Israelis, Arab nations announced an oil embargo. The price of oil soared, and lengthy lines queued up at gas stations. In such a climate, the pipeline's opponents stood little chance. On November 12, the U.S. House of Representatives passed the Trans-Alaska Pipeline Authorization Act of 1973 by 361 to 14; the next day, the bill passed the Senate by a vote of 80 to 5.

On April 29, 1974, using a patched-up Hickel Highway to transport equipment to the slope, Alyeska began construction on the Haul Road. The first length of pipe was laid on March 27, 1975, and the final weld completed on May 31, 1977. Just more than two months later, on August 1, 1977, the tanker *ARCO Juneau* left Valdez with the very first load of North Slope crude.

* * *

At the time of this writing, oil has been flowing from Prudhoe Bay, far above the Arctic Circle, to Prince William Sound via the Trans-Alaska Pipeline for twenty-three years. And at first glance, environmentalists' fears would seem to have proven groundless. Alaska has not disappeared beneath a morass of pipeline and pavement. Its skies have, mostly, not been darkened by pollution; its rivers and coasts appear relatively unspoiled. Throughout the vast majority of its considerable extent, the Great Land remains seemingly pristine, its breathtaking backdrops of mountains, glaciers, forests, and tundra filling residents with pride and awe and drawing more than 2 million visitors annually. But initial appearances can be deceiving, and the arrival of the oil industry has not by any means been without consequences.

In the region of the oil fields, more than 1,500 miles of roads and pipelines and thousands of acres of industrial developments cover 1,000 square miles of once-pristine tundra. Oil facilities on the North Slope emit as many polluting nitrogen oxides (contributors to smog and acid rain) as do the states of Vermont and Rhode Island and twice as much as does Washington, D.C. On the North Slope and along the pipeline, there is roughly one spill—of oil, diesel fuel, acids, or other materials—every day: there were 1,600 of them, involving 1.2 million gallons, between 1994 and 1999 alone. A draft 1988 report by the U.S. Fish & Wildlife Service—leaked to the *New York Times* but, controversially, never officially released in final form—found that construction and development had destroyed twice as much vegetation as originally predicted and that populations of most bird species had declined in the area, as had numbers of bears, wolves, and other predators. As predicted, the pipeline has essentially split the Central Arctic caribou herd in two, with females abandoning concentrated calving areas that have been encroached on by oil development. In addition, since 1995, the herd has declined in size from approximately 23,000 to roughly 18,000 animals—although that is actually a higher number than before TAPS construction began and reflects a cycle of increase and decline that has also occurred in caribou herds where there is no oil development. (A 2001 study shows that the herd has rebounded to 27,000 animals.)

Oil is now the source of 60 percent of Alaska state revenues, and it remains the fuel that drives the engine of the global economy—as well as countless cars, trucks, and sport utility vehicles across the country and around the world—and so, more than thirty years after the first strike at Prudhoe, exploration and extraction continue along the North Slope. In fact, the search for new oil is driving exploration still farther north, toward the fringes of the Arctic Ocean. In late 2001, BP expects to begin production from its Northstar project in the Beaufort Sea, the first truly offshore oil development in the Arctic, from which oil will be transported to land via subsea pipeline. That last fact in particular worries environmentalists, who charge that the pipeline, which will be buried in permafrost just six to eight feet below the surface of the seabed, will be vulnerable to scouring from ice floes, raising the prospect of leaks or a spill. They note that an assessment by the U.S. Army Corps of Engineers concedes as high as a one in four possibility of a major spill occurring during the lifetime of the project. The corps acknowledges, too, that under the worst of conditions—during the dark Arctic winter, with fierce winds, snow squalls, and ice-covered seas—responding to such a spill will be almost impossible.

Such is the concern over Northstar that during BP's annual general meeting in London in April 2000, 13 percent of proxy shareholders responded to a campaign by environmentalists by voting for a resolution calling for the project to be halted. Even so, the development continues, and meanwhile BP and other oil companies march on, expanding their range to include other sites in the Beaufort Sea and a corner of the National Petroleum Reserve in northwestern Alaska, and avidly eyeing the opportunity to drill for oil in the Arctic National Wildlife Refuge.

The Arctic National Wildlife Refuge comprises nearly 20 million acres of mountains, tundra, boreal forest, shrub thickets, and wetlands in northeastern Alaska, stretching from the Brooks Range to the coast of the Beaufort Sea and contiguous with Canada's Ivvavik National Park (formerly Northern Yukon National Park) to the east, the two combining to form what has been called the "largest protected block of wild habitat in the world." First set aside in 1960 as an Arctic National Wildlife Range, the refuge more than doubled in size in 1980 and is now the largest wildlife refuge in the United States—so large, in fact, that South Carolina could almost fit inside its borders. It contains the greatest variety of plant and animal life of any conservation area in the circumpolar north, including 36 species of fish, 180 bird species from four continents, and 36 species of land mammal, among them all three species of North American bear: black, brown, and polar.

Every summer, the 150,000-strong Porcupine caribou herd migrates en masse to its calving grounds in the refuge's coastal plain in a spectacle so dramatic it has earned the region the epithet of "America's Serengeti." Although the coastal plain constitutes less than 10 percent of the total area of the refuge, during summer it is the biological heart of the region. However, whereas 8 million of the refuge's 19 million acres are classified as wilderness, which means that no roads, mining, or development of any kind can take place there, only 30 miles of the 125-mile coastal plan are so designated. That is significant because according to studies by the U.S. Geological Survey, beneath the coastal plain there may lie anything from 5.7 billion to 16 billion barrels of crude oil.

The oil industry has long lobbied hard to be allowed to drill in the coastal plain. In 1995, Congress added a rider to the federal budget bill that assumed revenues from oil leasing in the area. Citing, among other things, objections to drilling in the Arctic National Wildlife Refuge, President Bill Clinton vetoed the budget, the federal government shut down, and the White House operated with a skeleton staff, during which time a young intern delivered a slice of pizza to the president in the Oval Office and history was made. In 2000, with oil prices rising and elections looming, cries from the industry and its congressional allies for drilling to be allowed increased still further. It is an argument that seems certain to grind on for some time to come. In a rehash of the claims made on behalf of the Trans-Alaska Pipeline, drilling's advocates insist that oil from the refuge is essential for issues of national security and self-sufficiency and that any environmental damage would be minimal. In response, environmentalists argue now, as they did during the debate over the pipeline's authorization, that the amount of oil that could be extracted would never be enough to reduce the country's reliance on foreign sources and that a better path would be to improve energy conservation and thus reduce oil use overall. And to assurances from the industry that there is no cause to fear environmental damage, environmentalists snort derisively. That, they point out, is a claim they have heard before.

The Day the Water Died

Even those who, during the pipeline authorization hearings, dismissed the natural worth of the Alaskan Arctic did not attempt to belittle the beauty of Prince William Sound, at the pipeline's southern end, or to deny the devastation that would be caused in the event of a catastrophic oil spill. They argued instead that such a spill would be impossible, that the safety and response systems at Valdez would be among the best in the world. But a 1989 committee established by the mayor of Valdez found that although ships sailing in and out of the port accounted for 13 percent of the country's tanker traffic, they were responsible for 52 percent of its accidents. Most of these were minor: before that review, however, and since, several close shaves have come perilously close to being disasters—and one accident proved every bit as devastating as anything in environmentalists' worst nightmares.

In 1980, as tugs stood by helplessly, the tanker *Prince William Sound* lost power in seventy-knot winds. For seventeen hours, the fully laden vessel was out of control in the middle of the sound; it was minutes away from wrecking on a reef when the winds changed, pushing it back into deeper water, and the captain was able to restore power to the engines. In 1994, the *Overseas Ohio* struck an iceberg, which tore a twenty-foot gash in the hull; disaster was averted only because the ship was heading into Valdez and so was not yet carrying oil. In 1995, the tanker *Kenai* veered out of the shipping lane and came within half a ship's length of wrecking on a shoal.

In January 1989, workers in San Francisco noticed that the *Thompson Pass,* a BP tanker scheduled to depart for Alaska, was leaking oil. Even so, company officials elected to send it north, where it promptly spilled 72,000 gallons into the Port of Valdez. It took Alyeska workers fifteen days to contain the spill, but the U.S. Coast Guard termed the response "adequate," and Alyeska considered the cleanup "textbook." Two months later, on March 23, Alyeska technicians and executives gathered in the Valdez civic center for a "safety dinner," congratulating themselves on having corralled almost all the oil from the largest tanker spill in Valdez thus far.

Not everyone was quite so impressed. As the Alyeska celebrants wined and dined one another, Valdez mayor John Devens was hosting a town meeting to hear residents' concerns about the possible consequences of oil spills in Prince William Sound. Alyeska declined an invitation; Riki Ott, however, did not. Ott, a fisherman and biologist from Cordova, had planned to attend the meeting, but fog prevented her plane from leaving. Instead, she was patched through by teleconference and spoke at length about the pollution problems, accidents, and close calls that were already a factor of tanker traffic in the sound, and the inevitability, as she saw it, of a much more significant spill occurring in the near future. "We're playing Russian roulette," she warned. "It's not a matter of *if.* It's just a matter of *when* we get the big one."

The meeting ended about 9:00 P.M. At 9:12, the oil tanker *Exxon Valdez,* carrying nearly 1.5 million barrels (around 62 million gallons) of Alaska crude, moved gently away from the dock. Pilot Ed Murphy then began steering the vessel out of the harbor, through the Valdez Narrows, and into the waters of Prince William Sound.

At 11:24 P.M., two hours later and fourteen miles from Valdez, Murphy disembarked and command reverted to Captain Joseph Hazelwood. The ship began to accelerate to sea speed, around eleven knots, but ahead, radar showed that a thick plume of small icebergs, jettisoned by the retreating Columbia Glacier, was blocking its path. Steering to starboard to skirt the obstacle was not possible: that was the direction of the glacier and therefore the direction from which the ice was streaming. Hazelwood could stop the ship and wait for the ice to clear, or he could slow the tanker and thread through the floes gently. Or he could exercise another option: turn to port and steer around the leading edge of the ice before returning back on course. The latter was the option Hazelwood chose, but it was not without risk. For one thing, shipping in Prince William Sound, like traffic on a freeway, is consigned separate lanes for incoming and outgoing vessels; Hazelwood's chosen maneuver would entail him crossing into the lane for ships entering the harbor. Of greater moment, however, was that it would take the *Exxon Valdez* perilously close to dangerous Bligh Reef. Such was the size of the tanker that it would require six-tenths of a mile to make the sharp turn between the reef and the ice; the gap itself was only nine-tenths of a mile in width.

At 11:53 P.M., Hazelwood confirmed the order with Third Mate Gregory Cousins and stepped down from the bridge. At 11:55 P.M., the lookout advised Cousins that a red navigational light on the Bligh Reef buoy, which should have been ahead and to port, was flashing to starboard. Cousins ordered the helmsman to make a ten-degree right turn. Either the helmsman or the vessel itself was slow to respond, and several minutes later the light was still on the starboard side. Cousins ordered the ten-degree turn again, then a twenty-degree turn, then a hard right. It was too late.

He called Hazelwood. "I think," he said, "we are in serious trouble."

At 12:04 A.M., the *Exxon Valdez* struck Bligh Reef. The ship shuddered and shook, scraped along for a further 600 feet, and ground to a halt.

* * *

At 12:27 A.M., Hazelwood called Coast Guard traffic control in Valdez. "We've fetched up," he announced, "run aground north of Goose Island, around Bligh Reef. . . . And we're going to be here awhile."

There was more. "Evidently," he continued, "we're leaking some oil."

Indeed. The ship had lost about 100,000 barrels—more than 4 million gallons—in just half an hour. Oil from one of the tanks was shooting forty or fifty feet into the air. By the time a Coast Guard team reached the tanker, around 3:00 A.M., the oil was 12 to 18 inches deep against the hull. Within five hours of the grounding, an estimated 240,000 barrels—about 11 million gallons—of crude oil had escaped from the *Exxon Valdez* and into Prince William Sound.

* * *

At the time of the pipeline's authorization and for the first few years of its operation, Alyeska had committed to and maintained a dedicated oil spill response unit. In 1982, the consortium disbanded that team, arguing that its duties could be performed just as easily by other personnel and that the right people and equipment would always be ready in the event of an accident. The *Exxon Valdez* put that contention sorely to the test.

Alyeska's contingency plan stated that a vessel with containment equipment would arrive at the scene of a spill within a maximum of five and a half hours. At 5:00 A.M.—five hours after the *Exxon Valdez* ran aground—Alyeska employees waited at the terminal, expecting to be given assignments. But the equipment necessary to contain the spill was not on the barge that was supposed to transport it there. In fact, the barge itself was in dry dock awaiting repairs for a crack in the hull sustained during a storm in February. Alyeska decided that crack or no crack, the barge was seaworthy enough to respond to the spill, but meanwhile the needed equipment was buried in a warehouse. It was a good eighteen hours after the accident before the first booms were deployed, and even then the number and size of those booms were insufficient to contain the slick.

For several days, the spill went largely uncontained. Exxon and Alyeska squabbled over which entity was responsible for doing what; partly as a result, neither did much of anything. As the citizenry in Valdez and Cordova grew anxious and angry, the slick spread slowly across the waters of Prince William Sound.

On day two of the spill, Exxon tested the use of dispersants—chemicals meant to dilute the oil by causing it to mix with seawater—but their effectiveness was inconclusive, their use controversial, and anyway, there just was not enough available to tackle a spill of that size. The company also tried burning off a small patch of the oil to see whether that would have any effect, but it filled the sky with thick, acrid, choking smoke.

By the time the response effort finally was fully mobilized, with booms and skimmers ready to be deployed and enough people on the scene to make the operation work, it was too late. Overnight between Sunday, March 26, and Monday, March 27, a nascent storm high in the Chugach Mountains gathered strength, and when Monday dawned, fifty-knot winds bore down on Prince William Sound. In a matter of hours, the oil slick, which had until then been a little less than four miles in length, grew to forty miles long. It continued to expand, spreading south and west. When the winds abated two days later, the slick encompassed hundreds of square miles, and seemingly every spare inch of beach, bay, and inlet along western Prince William Sound was covered with thick, tarlike oil. Still the slick kept spreading—to Kodiak Island, to the lower reaches of Cook Inlet, and out into the Gulf of Alaska—until ultimately, weeks later, it had washed up along almost 700 miles of shoreline, as far as 470 miles from the initial spill site.

It was the start of the spring migration, and millions of seabirds were beginning to descend on the sound. Around the world, television cameras broadcast the inevitable result: birds landing on the water or on the beach and in moments becoming entombed in thick, black oil, then struggling to right themselves and escape but being dragged down in the slick's viscous embrace. Bald eagles swooped down on the dead and dying birds and themselves fell victim to the crude's poisons. Sea otters washed up on the shore, their teeth bared in agony, their fur soaked with oil.

The spill killed an estimated 250 bald eagles, 300 harbor seals, and 3,500 to 5,000 sea otters. Fourteen of thirty-six members of the resident "AB" pod of killer whales died within three years—an unprecedented rate of mortality for North Pacific orcas.

More than 35,000 carcasses of oiled birds were recovered in the first four months, and overall estimates of the number of seabirds killed range from 300,000 to 675,000. The spill reduced the area's population of pigeon guille-mots by 10–15 percent and the population of common murres by as much as 40 percent. It may have killed as many as 2,000 Kittlitz's murrelets—a sub-stantial proportion of the world population of a species found only in Alaska and portions of the Russian Far East.

Less easy to quantify but no less real were the psychological and sociolog-ical scars the oil spill left on the communities around Prince William Sound. Many felt violated by the despoliation that had suddenly been visited on their corner of paradise; fishermen broke down in tears—great, heaving sobs—as they tried to rescue birds and otters from the oil's clutches. Those who earned their living from catching fish suddenly found their livelihood torn from them; even now, says Riki Ott, "no one can make a living fishing exclusively in the sound anymore." Communities were swamped by the sudden influx of cleanup crews, news media, rubbernecks, lawyers, and VIPs (a special viewing platform was built for Vice President Dan Quayle to survey the damage) and were torn apart by rifts between those who refused the oil companies' money and those who accepted sizable payments to contribute to cleanup efforts—rifts that in many cases have not healed.

Such consequences were in many cases even more acute for native com-munities, which depend on the sound for their subsistence way of life. "If the water is dead," wrote Walter Meganeck Sr. in the *Anchorage Daily News,* "maybe we are dead, our heritage, our tradition, our ways of life and living and relat-ing to nature and each other. . . . Never in the millennium of our tradition have we thought it possible for the water to die, but it's true."

Exxon launched a colossal cleanup campaign, spending at least $2.5 billion on cleansing shores and rehabilitating wildlife. Workers were paid $16.95 per hour to scrub rocks. Centers were set up for the cleaning and care of oiled wildlife. Birds and otters were washed and fed, and an estimated 250 otters were saved—at an approximate cost of $90,000 each. Critics charged that the effort was little more than an expensive, showy public relations stunt, and certainly some of the cleanup did more harm than good: high-pressure hoses blasted scalding water onto beaches, and although that certainly cleaned the rocks, it also steamed the beach life to death.

Despite the massive effort, scientists in 1992 calculated that just 14 percent of the oil had been recovered or destroyed through cleanup efforts; 70 percent had evaporated or degraded. Perhaps 1 percent had dispersed into the ocean, spreading out at extremely low concentrations into the Gulf of Alaska. Thirteen percent or so, however, remained in mud, sand, and gravel, and 2 percent could still be found on the beach at scattered locations. Since 1993, no one has attempted a comprehensive survey or calculated the volume of remaining oil.

✻ ✻ ✻

Prince William Sound, almost every researcher will say, has recovered astonishingly well from the obscenity of March 1989, and it continues to do so. It is, in most respects, a picture of health. Certainly, it stands as a profound tribute to the extraordinary recuperative powers of nature. But it is healthy in the sense that someone escaping from an addiction, a trauma, an abusive relationship, is healthy. It functions perfectly well, looks as beautiful as ever. But beneath the surface, there are scars.

More than ten years after the accident, says the *Exxon Valdez* Oil Spill Trustee Council—a scientific body established with reparations paid by Exxon—only two species can truly be said to have recovered from the spill: the bald eagle and the river otter. Although many are showing strong signs of improvement, others are not. The harbor seal population continues to decline by about 5 percent per year, although other environmental factors may be involved. Pigeon guillemot numbers remain depressed along oiled beaches. Harlequin ducks are declining in the western sound, and survival rates of adult females are significantly lower in oiled areas. Pink salmon stocks crashed shortly after the spill and have been slow to recover; likewise, Pacific herring are increasing only gradually, and recent studies have shown that even minute amounts of crude oil can affect developing fish embryos and the immune systems of adult fish.

On a warm summer's day, with sea otters alternately lying on their backs and disappearing beneath the water and bald eagles soaring overhead, there is no obvious sign that humanity has violated Prince William Sound in any way—no obvious sign, from some vantage points, that humanity has ever been there at all. It seems in many places unchanged and unchanging: looking, surely, almost exactly as it did when Captain Cook first found himself there more than two hundred years ago.

But the evidence is there, just below the top layer of rocks on some of the beaches that were directly in the spill's path: thick, black, and viscid, an ugly vestige of the stain that befouled paradise, a reminder of the day the water died.

Chapter 7

The Ends of the Earth

When Richard Byrd completed his pioneering flight to the South Pole in 1929, aviation was still in its infancy. Byrd's expedition was only the second to employ an airplane successfully in Antarctica, just one year after Hubert Wilkins had, and it came only two and a half years after the first solo flight across the Atlantic Ocean. Not until the 1950s and, particularly, activities in conjunction with Operation Deepfreeze and the International Geophysical Year did the use of aircraft truly become established on the frozen continent.

Five decades on from Byrd's achievement, several of the countries operating in Antarctica used airplanes to supply their bases; flights to McMurdo Station from the New Zealand town of Christchurch were commonplace, and the connection from the Ross Ice Shelf to the South Pole—an extraordinary and unprecedented achievement when accomplished by Byrd and comrades—was now completed regularly during the summer months to supply the United States' Amundsen-Scott South Pole Station.

But Wilkins and Byrd had been the pioneers, and it was in commemoration of the late admiral's achievements that some of Byrd's colleagues and relatives set out on board a U.S. Air Force C-141 Starlifter from Christchurch for McMurdo Station early on November 28, 1979, fifty years to the day since Byrd made that first South Polar flight.

On any other day, it might have been the stuff of news stories and human interest features, an opportunity to reflect on the growing success and safety of Antarctic aviation. But the C-141 was not the only airplane to set out that morning for Mactown, as McMurdo Station is colloquially known among Antarcticans, nor was it the first. At 8:17 A.M., Air New Zealand Flight 901 departed Auckland with 237 passengers and 20 crew members, and it was this flight that would dominate the headlines.

Unlike the party commemorating Byrd's flight, the people on board Flight 901 would not be staying at McMurdo or continuing on to the Pole; in fact, they did not intend to land at all. It was a purely sightseeing venture: a one-day excursion to Antarctica scheduled to fly over the subantarctic Auckland Islands, head for the Balleny Islands, and then turn to fly past Cape Adare, south over the Ross Sea, and around Ross Island before heading back to New Zealand in time for a late dinner.

The Australian airline Qantas Airways Limited had been the first to introduce these so-called champagne flights to Antarctica in 1977, but Air New Zealand had followed close behind; this was the fourteenth such flight the airline had run. The round trip was expected to take about twelve hours, including a stopover in Christchurch on the way north.

All the previous Air New Zealand flights to Ross Island and back had, programmed into the in-flight computer, a flight track that took them south along McMurdo Sound and comfortably west of Mount Erebus before swinging them south and then east of the mountain to head home. The night before Flight 901 took off, however, the airline's navigation department, in response to observations from pilots that the coordinates keyed in for Mactown were in error, altered the flight track. The new route took the plane much closer to the volcano; apparently, incredibly, nobody thought to inform the flight crew.

Disaster might have been averted had weather conditions been clear; in such circumstances Mount Erebus, a magnificent and imposing sight, is not easily missed. But this was Antarctica, where the weather has a long tradition of noncooperation with the plans of human interlopers, and on this day the approach to Ross Island was blanketed with thin, low clouds. As his destination neared, Captain Jim Collins had taken the airplane below the cloud layer, but with no sunshine reflecting off the snow below, there was no visual contrast. Collins was in a white-out, and just before the alarms of the ground proximity warning system resonated in the cockpit, he seems to have recognized as much. His instruments told him he was only thirty miles from McMurdo, but ahead of him he could see nothing: no Mactown, no mountains, no glaciers, no snow, no ice, no horizon—nothing at all.

"We'll have to climb out of this," he said, but it was too late. Seconds later, Air New Zealand Flight 901 flew straight into Mount Erebus. There were no survivors.

* * *

It is an extraordinary phenomenon, the way the trickery of the polar light can make an enormous mountain simply disappear, in the same way it can make that same mountain seem almost close enough to touch. One morning, I stood at Cape Evans and looked up at the tremendous sight of Mount Erebus in the distance, a plume of vapor rising from its crater, the phenomenon of fata morgana throwing the snow-shrouded volcano into sharp, dramatic relief. That afternoon, I looked up from the exact same spot, in the exact same direction, and saw nothing. There was no obvious cloud layer hiding the mountain from sight, no hint that my view was in some way being obscured or that with a change in the wind, the volcano would once more be revealed. Erebus had completely disappeared. There was simply nothing to see, just Ross Island continuing into the distance beneath a pale sky.

* * *

The demise of Flight 901 was a tragedy, an appalling loss of life and by far the worst single disaster ever to befall human activities in Antarctica. In New Zealand, it also spawned controversy and criticism of the airline, which stood accused of attempting to minimize its own complicity and seeking instead to blame human error on the part of the pilot. In addition, it focused attention on an issue that few had considered before: that of tourism to the Antarctic and subantarctic regions. Few indications that a wilderness has been tamed are as vivid as that wilderness' regular appearance on tourist itineraries; one revelation of the Erebus disaster was that less than seventy years after Antarctica's conditions had thwarted Ernest Shackleton and killed Robert Falcon Scott, it was possible to have an early breakfast, fly to Ross Island, and be back by bedtime.

Getting Away from It All

In his authoritative chronology of Antarctic expeditions, Bob Headland of the Scott Polar Research Institute of the University of Cambridge says that although details are difficult to confirm, New Zealand government expeditions to islands on the fringes of the Antarctic may have carried tourists as early as 1882. An 1891 visit to the subantarctic Macquarie Island by the ketch *Gratitude* also seems to have been at least partly a tourist junket. By 1910, the travel agency Thomas Cook & Son was planning an expedition to the Ross Sea; it was reported that "members of the New Zealand parliament, a number of ladies and several gentlemen interested in scientific matters" were keen on the enterprise, but nothing came of it. Passenger spaces were available on mail ships to South Georgia Island in the 1920s, and on the Argentine transport ship *Pampa,* resupplying the meteorological station on Laurie Island, in 1933. The first tourist flight to Antarctica was on December 22, 1956, when a LanChile Airlines Douglas DC-6B flew from Chabunco, Chile, and back on a tour of Chilean bases in the Antarctic Peninsula region. The next season, Argentina responded with seaborne tourism: two tours, of nine and twelve days' duration, organized by the Naval Transport Command.

The man who truly breathed life into the notion of Antarctic tourism, however, was Swedish adventurer Lars-Eric Lindblad. As a young boy growing up in prewar Stockholm, Lindblad had been consumed with the idea of journeying to lands beyond, especially Antarctica: "I didn't just read about Amundsen, Shackleton and Scott," he wrote, "I felt I was *there* with them." In 1947, fresh out of college, he took a job with Thomas Cook & Son before electing to move to the United States, where by 1952 he had landed a job in the New York office of the Netherlands-based travel agency Lissone-Lindeman. Although his work allowed him to indulge, at least vicariously, his lifelong love of globe-trotting, he felt frustrated. Most tours of that time, he recalled,

> were concentrating on only a tired handful of countries, or Grand Tours of Europe. . . . In the meantime, the more beautiful and fascinating places of the world—the Middle East, the African wilds, the islands of the Pacific, India, Japan—were largely being ignored. My earliest dreams had been of these

places; I was burning with the desire to go to them. Travelers, I thought, must be feeling the same way.

By his second year at Lissone-Lindeman, he had begun to receive "a trickle of requests from more adventurous travelers" who "wanted to reach out to more exciting places" and, more important, could afford to do so. His enthusiasm piqued, he set out on a kind of exploratory tour of the world's more remote regions, and by 1958 he had founded his own company, Lindblad Expeditions. He organized tours for the Garden Clubs of America to see the "botanical and horticultural wonders of the world" and for the Archives of American Art to view the artistic antiquities of Egypt and Greece. He led safari adventures to sub-Saharan Africa and negotiated his way through the iron curtain to take American vacationers to Siberia and Mongolia. Ultimately, Lindblad Expeditions would expand across the globe, exposing curious, wealthy westerners to such people and places as pygmies in Zaire, dancers in Bali, and monks in Bhutan. But the polar regions remained the true love of the company's founder.

Lindblad chartered an Argentine naval transport vessel, the MS *Lapataia*, and in 1964 took his first passengers on a journey from Ushuaia, Argentina, to the Antarctic Peninsula. Despite a few inevitable teething troubles, recounted Lindblad, "the first tentative probe into Antarctica was a success"—so much so that by the time he returned to New York, he wrote, "we were flooded with inquiries about our next such venture." Lindblad was even invited to a meeting with Chile's president, Eduardo Frei, who, having noted that the previous voyage had been on an Argentine-flagged ship, offered the use of the Chilean vessel *Navarino*. During the 1966–1967 season, Lindblad organized three tours, one on the *Navarino* and two on the *Lapataia;* commercial cruise ships have visited the Antarctic every summer since.

* * *

That first cruise carried 56 paying customers; subsequently, particularly during the 1990s, the number of tourists visiting Antarctica by ship increased dramatically. According to the International Association of Antarctica Tour Operators (IAATO), 106 expeditions carried 9,604 people to the Antarctic during the 1997–1998 season; the following year, 10,013 tourists traveled south on 116

trips. As I write this, final figures are not in for the 1999–2000 season, but
IAATO estimates project another increase, to as many as 143 voyages and
almost 16,000 tourists—a lot of people visiting a fragile environment.

Defenders of Antarctic tourism point out that it in effect "democratizes" the
continent and makes it accessible to people beyond the scientific and political
elite. After all, why have a de facto World Park if the world is not allowed to
visit it? In addition, Antarctica, perhaps more than any other place, has a deep,
even spiritual effect on a great many who have been fortunate enough to go
there; expanding the circle of those able to enjoy that experience can only
increase the constituency of supporters for Antarctica's continued wilderness
status. As one experienced Antarctica hand has put it, it is "important that lots
of us go to the Antarctic to enjoy and wonder at the sheer beauty and abundance of the place, because that will produce a body of enlightened ambassa-
dors, our best guarantee that this mighty wilderness will endure." Tourists were
in fact among the earliest and most vocal critics of the appalling environmental conditions around many scientific stations, and even the *Gratitude*'s excursion to Macquarie Island in 1891 spawned public censure of the hunting of seals
and penguins there. Many within national scientific programs also appreciate
the fact that taxpayers can see their work in action and, it is hoped, return
home to support the programs' continued funding. But the sheer number of
people now visiting Antarctica—and even more, the fact that many of those
visitors are disgorged on the continent several hundred at a time from large
cruise ships—is beginning to have other, negative effects.

Long before thoughts turned to the development of an Environmental Protocol for the Antarctic, Lars-Eric Lindblad cautioned that he wanted travelers
on his tours to Antarctica to leave nothing but footprints: a laudable goal, but
in Antarctica even a footprint—if left in a moss bed, for example—can be
damaging. Furthermore, the vast majority of the footprints left in the Antarctic, and almost all the tourists who leave them, occur in a select few areas of
the continent: essentially, the Antarctic Peninsula region and Ross Island. These
areas are attractive to tourists for the same reasons they are home to so many
scientific bases: they are, at least in places, free of ice year-round, and they contain interesting concentrations of wildlife. (In addition, of course, the very
presence of bases, and of historic huts and shelters, is itself a draw for tourists.)

But those same attractive features make them especially vulnerable to disturbance.

Even standing and watching wildlife from an apparently safe distance can have an adverse environmental effect: one study showed that the heart rate of incubating Adélie penguins increased markedly when the penguins were approached by a human who was still 100 feet away. The researchers who conducted that study also found that a person standing 66 feet from a well-used penguin "pathway" to the sea caused the penguins to deviate from their path by almost 230 feet, even several hours after that person had left. In one case, a single observer caused almost 12,000 penguins to take a longer route to sea. And although penguins taking a detour to their destination might not seem like much of an adverse effect, in the harsh Antarctic environment every iota of energy is hard to come by and can ill afford to be squandered.

Few if any environmental groups advocate a ban on Antarctic tourism; IAATO points out that several of its tours are run in conjunction with such organizations as the Smithsonian Institution and the World Wildlife Fund. IAATO has also adopted a stringent code of conduct for its customers, much of which has been incorporated into Antarctic Treaty guidelines. Nonetheless, even if all visitors to Antarctica strictly abided by the code, their growing numbers would still be cause for concern.

Already, there are signs that the mere presence of humans has begun to alter the Antarctic environment. Species of non-native grass, presumably carried in on the clothing of visitors or scientists, have been found at several sites in the Antarctic Peninsula region and on the continent itself, and Australian researchers announced in 1997 that populations of emperor and Adélie penguins had been found to have antibodies to a poultry virus that had almost certainly been introduced by humans.*

---

*Six years earlier, scientists discovered canine distemper antibodies in crabeater seals. For environmentalists, the find justified their successful campaign to have sled dogs banned from the Antarctic under the Madrid Protocol. Many of those who live and work on the frozen continent, however, most of whom are otherwise highly supportive of conservation moves in the region, absolutely hate the ban on dogs, depriving them as it does of much-appreciated canine companionship, especially over the winter months.

With the advent of the environmental protocol, tourist activities in Antarctica may come under yet tighter control. Some treaty members are anxious to limit tourist numbers or to otherwise apply to tourism the same kind of restrictions and oversight that theoretically govern official activities. So far, however, talk of formally incorporating a tourism annex to the protocol—a move that would give the existing guidelines some teeth—has come to naught.

* * *

Barely a generation ago, Antarctica remained almost completely unexplored beyond coastal areas and the path from the Ross Sea to the South Pole. Less than two hundred years ago, nobody knew for sure that it even existed. Now, anyone with the desire and the finances can book passage on a cruise ship and dine in comfort while looking through enormous windows at ice-strewn waters that horrified the likes of Captain James Cook.

In the Arctic, the story is similar. Soviet aviator P. A. Gordienko, the first person to stand, beyond all doubt, at the North Pole, accomplished his feat as recently as 1948; assuming—as many authorities now do—that the claims of both Frederick Cook and Robert Peary were false, nobody reached the Pole by crossing the ice of the Arctic Ocean until Ralph Plaisted did so in 1968. Since then, Ward Hunt Island—just off the northern coast of Ellesmere Island and the base camp for most polar assaults—has become littered with discarded equipment and trash from an array of expeditions to the North Pole. Less adventurous souls can ride on an icebreaker and, when their ship reaches its destination, clamber down to the ice for a brief walk around the top of the world.

Arctic tourism is on an altogether different scale from that in Antarctica and most likely will remain so in perpetuity, no matter how much visitors to the Antarctic grow in number. The Arctic's climate, although harsh in parts and at times, is milder than that of the southern continent, and the region is far closer to major population centers. These factors make it more accessible and enable visitors to indulge in a greater array of comforts than those on offer in the south. Although precise estimates vary, tourists to the Arctic and subarctic are generally reckoned to number 2 million to 3 million annually, substantially more than the 15,000 or so who see the Antarctic each year.

Whereas tourists in the Antarctic are largely confined to cruise ships, visitors to the Arctic need not be subject to such restrictions. If they choose to base their excursions out of a town such as Tromsø, in northern Norway—home of, among other things, the northernmost planetarium, university, eighteen-hole golf course, and Burger King restaurant in the world—or a subarctic city such as Anchorage, they can delight in all the comforts of home and then some, staying in a five-star hotel, enjoying a late-evening stroll down a sun-drenched street, taking in a movie, and finishing off the night with drinks at a bar.

As with tourism, so with most other aspects of human activity in the polar regions. At the dawn of the twenty-first century, both remain among the most remote, inaccessible, inhospitable areas on Earth. But whereas the Antarctic continues to stand largely apart from the rest of the planet, the modern world has made its presence felt, for better and for worse, throughout much of the Arctic. Indeed, given the image of the Arctic that persists in the popular consciousness—namely, that of a vast, freezing wilderness, a perception that remains valid for vast swaths of the region—the frequency and extent of the industrialized world's presence is perhaps surprising.

The Atomic Age in the Arctic

The largest nuclear device ever exploded was not at Bikini Atoll, nor in Nevada, nor at the French testing site of Moruroa Atoll in the South Pacific, but in the Arctic. In October 1961, the USSR detonated a fifty-eight-megaton thermonuclear bomb—roughly six thousand times more powerful than the weapon dropped on Hiroshima—on Novaya Zemlya, off the northwestern coast of Russia. The blast generated radioactive fallout that was detected as far away as Alaska and Japan. All told, the Soviets conducted 120 nuclear tests on Novaya Zemlya, three-quarters of them atmospheric tests, which were all conducted before completion of the Partial Test Ban Treaty in 1963. About 12 percent of the fallout from those tests—as with tests conducted elsewhere in the world—was deposited in the immediate vicinity of the site, and another 10 percent entered a band circling the planet at about the same latitude. The rest was picked up by winds and redistributed, along with pollution from all the other atmospheric tests conducted by the nuclear powers, around the globe. According to a report by the Arctic Monitoring and Assessment Programme

(AMAP), the net result of this fallout is likely to be a total of 750 additional cases of fatal cancer throughout the Arctic.

Novaya Zemlya was not the only place in the Arctic and subarctic to host nuclear weapons tests. On three occasions—in 1965, 1969, and 1971—the United States conducted underground tests on Amchitka Island in the Aleutian Islands chain; the third in the series, code-named Project Cannikin, was the largest nuclear explosion in U.S. military history. The tests met with massive public opposition, partly because emotions had been heightened by the Vietnam War, but partly also because of their location. The region is crisscrossed with major fault lines, and on March 27, 1964, Alaska had been battered by the second largest earthquake ever recorded: a magnitude 9.2 temblor centered beneath Prince William Sound that flattened large parts of Valdez and Anchorage and caused tsunamis to crash on the shores of Oregon, California, Hawaii, and Japan. Fearing a recurrence, 10,000 protestors blocked the border crossings between the United States and Canada prior to the 1969 explosion: "Don't Make a Wave," the demonstrators' signs read; "It's Your Fault if Our Fault Goes." In the end, there was no earthquake and there were no tsunamis; but concern persisted, and one group of activists greeted the announcement of the 1971 test by setting out for Amchitka in an old halibut seiner. Although their plans to "bear witness" to the test were thwarted by the weather and the attentions of the U.S. Coast Guard, the legacy of their effort would endure; for reasons that remain obscure, the protestors called themselves "Greenpeace," and by the time they returned to shore, a movement had been born.

Nuclear devices were also exploded in the Arctic for reasons other than to test their adeptness at annihilating the enemy. The Soviets used them for all kinds of things: to construct dams, to carve out canals, to redirect rivers. Not surprisingly, at least some of the projects went awry and released radioactivity into the environment; almost as unsurprising, given the Soviet state's penchant for secrecy and paranoia, is the limited information available about their effects. Contamination levels from a 1974 explosion were not measured until 1990, and the official report of the investigation gives few details beyond referring ominously to a "dead forest."

Similarly, in 1958 the U.S. government announced plans to detonate as many as six thermonuclear weapons off Cape Lisburne, Alaska, north of the

Arctic Circle, in order to create a new harbor. The plan was the brainchild of Edmund Teller, the self-styled father of the H-bomb; dubbed Project Chariot, it was part of a wider effort called Project Plowshare, which sought to concoct ways in which nuclear weapons could be put to nonmilitary use.

Although Teller and others tried to sell the alleged commercial benefits of a deepwater harbor in northwestern Alaska, it was essentially an experiment: a test of the potential for use of thermonuclear weapons in planned projects such as carving an alternative to the Panama Canal through Colombia, seeding rain clouds to change the weather, damming the Mediterranean Sea to flood and irrigate the Sahara, warming the Arctic, and, as Dan O'Neill pointed out in his excellent book *The Firecracker Boys,* generally improving "a slightly flawed planet." It was also, covertly, a test of the effects of radioactive fallout on the region's biota—including its Iñupiat inhabitants.

Although much of the territory's business community and even the University of Alaska bought into Project Chariot, a grassroots effort spearheaded by the Iñupiat of Point Hope and some maverick University of Alaska researchers was able to thwart it. Nevertheless, the project site was used as a dump for radioactive waste, and Point Hope villagers remain concerned about the possible link between the program and an apparently abnormally high cancer rate in the village.

Radioactivity has seeped into, and rained down on, the Arctic from far and wide. In 1968, a U.S. Air Force B-52 carrying four nuclear weapons crashed on the ice near Thule, Greenland. The impact triggered conventional explosives on board, which fragmented the nuclear weapons and scattered plutonium over the ice. Ten years later, the nuclear-powered satellite *Cosmos 954* tumbled back to Earth, spreading radioactive debris over 62,000 square miles of Canada's Northwest Territories; some of the debris was so radioactive that it had to be handled with long tongs by men standing behind a 1,600-pound lead shield. In 1989, the Soviet nuclear submarine *Komsomolets* caught fire and sank in the Norwegian Sea southwest of Bjørnøya; eleven years later, to the horror of people around the world, another Russian nuclear submarine, the *Kursk,* sank to the bottom of the Barents Sea, taking the lives of all on board. For more than thirty years, the USSR used the Kara Sea and the fjords of Novaya Zemlya as dump sites for radioactive waste, including nuclear submarine reactors.

Perhaps most famously, the 1986 Chernobyl explosion in Ukraine spread radioactive fallout north and west and ultimately to the Scandinavian Arctic, where it was absorbed by lichens, which in turn were eaten by reindeer. Reindeer are the traditional food of the Saami, an indigenous people of northern Norway, Sweden, Finland, and parts of Russia's Kola Peninsula. For several years after the Chernobyl accident, the body burdens of radioactivity more than doubled among Finnish Saami and increased tenfold among Saami in Norway and Sweden. Radioactivity has since mostly returned to pre-Chernobyl levels, but more than a decade later some Saami villages still give their reindeer special food to help the animals' bodies rid themselves of radiocesium.

The Arctic is in many respects more vulnerable to radioactive contamination than much of the rest of the world. Ocean currents transport radioactive material—such as discharges from reprocessing facilities—from Europe along the Norwegian coast and into the Arctic Ocean basin, where it can remain for years. Radioactive material trapped in ice is cleansed from the environment more slowly, and glaciers thus become reservoirs of old fallout. This is a major problem for communities such as those in northern Greenland, where the majority of drinking water comes from ice and snow and where levels of strontium 90 from atmospheric fallout are declining more slowly than on the southern part of the island. And the traditional diet of some polar peoples, particularly those who eat large amounts of caribou and reindeer meat, can accumulate especially high levels of radioactivity. All told, exposure to radioactive contamination is about five times higher for people in the Arctic and subarctic than for those in temperate regions; for those who eat large amounts of caribou and reindeer, it is some fifty times higher again.

War, Hot and Cold

On the night of June 2, 1942, the Japanese aircraft carriers *Junyo* and *Ryujo*, carrying eighty-two planes and accompanied by two cruisers and three destroyers, steamed across the North Pacific Ocean toward the Aleutian Islands. The Japanese invaded and occupied the islands of Attu, imprisoning its population of thirty-nine Aleuts and the schoolteacher, a Mrs. Jones, and Kiska, where the occupying force met no resistance from the ten members of a temporary U.S. weather station. In May 1943, after three weeks of fighting, American troops recaptured Attu. For the Japanese in particular, it was an extraordinarily bloody battle: from a force of around 2,600 men when the American action began on May 11, fewer than 800 fighting men remained by the end of the month. Of those, some 600 were wounded; on the evening of May 28, 400 were given lethal shots of morphine by their own officers. Only twenty-eight Japanese were captured; every other Japanese soldier who was not killed in battle committed suicide.

After Attu, the United States turned its attention to Kiska, bombarding it from sea and sky from June 1 to August 15. But the Japanese had decided to remove the garrison, and under the cover of darkness on the nights of July 28 and July 29, they evacuated the entire occupation force of more than 5,000 on board eight ships.

It is perhaps surprising, the extent to which battle has been joined by various countries in the polar and subpolar realms: skirmishes between British and Dutch over whales off Spitsbergen in the seventeenth century; war between Britain and Argentina over the Falkland Islands and South Georgia in the twentieth; the use by Nazi Germany of islands in the Antarctic and subantarctic as bases for its submarines. The fact that American territory on the fringes of the Arctic was actually invaded and, albeit temporarily, under enemy control is even less widely known. In fact, the Arctic and subarctic saw a great deal of action during World War II.

The United States used bases on Greenland as staging posts to fly warplanes to Britain, and convoys of Allied ships threaded their way through the icy waters north of Norway, even as far north as Bjørnøya, in an attempt to bring supplies to besieged Murmansk and Arkhangel'sk. The Barents Sea saw fierce fighting as the convoys came under attack from Nazi ships and aircraft; entire crews became shipwrecked on the frozen shores on Svalbard (the archipelago formerly known as Spitsbergen) and Novaya Zemlya.

With the end of World War II, wartime allies became peacetime adversaries, and the descent of a "Bering wall" bisected the strait between Alaska and Siberia. Natives on each side of the divide—relatives who for centuries had sailed and rowed back and forth without hindrance—were prohibited from visiting each other for almost fifty years. On Big Diomede Island, Soviet authorities forcibly removed the villagers and assimilated them on the mainland; for decades, one of the favorite sports for the youth on Little Diomede Island was racing across the water or sea ice between the two islands and trying to touch Big Diomede before the Soviet soldiers stationed there opened fire.

Nowhere else did East and West abut each other as directly and on as many fronts as in the Arctic: in the Bering Strait; on the border between Russia and Scandinavia; even directly across the polar ice cap. Each side used nuclear submarines to probe the other's Arctic waters, testing its opponent's alertness and resolve. The USSR turned the Kola Peninsula into a cold war fortress bristling with military bases and listening devices. From Greenland west across Canada and Arctic Alaska, the North Atlantic Treaty Organization set up a string of distant early warning (DEW) radar sites to warn of impending missile attacks.

Many of these DEW sites and associated military bases are now abandoned, leaving behind a legacy of pollution. In the vicinity of just one old communications and surveillance base on the Northeast Cape of St. Lawrence Island, the U.S. Army Corps of Engineers has identified twenty-three contaminated sites that require environmental investigation and cleanup. There are more than 220,000 gallons of spilled fuel and a dump that contains almost 30,000 barrels of toxic waste. Large bales of wire left on the tundra have trapped reindeer by their antlers; unable to escape, the animals have starved to death. In recent years, sixteen people have died of cancer in the St. Lawrence village of Savoonga—an extraordinarily high number for a population of just a few hundred—and other cases of cancer have been diagnosed. There have also been birth defects and premature births, which many in the village attribute to the toxicity of the site. Elders of the village point to a stream that flows north through the coastal tundra and into the Bering Sea; they say it used to be one of the richest fish streams on the island, but no fish have been there in more than thirty years.

Pollution, from Within and Without

Few areas of the Arctic can be said to be truly industrialized, but there are exceptions. Parts of the Russian Arctic in particular are subject to industries as capable of pollution as those found anywhere. According to a report by AMAP, the metal smelter at Noril'sk, in northwestern Siberia, belches around 1 million tons of sulfur into the air each year. At the border between Finnish Lapland and the Kola Peninsula, the report notes, "the coniferous forest appears healthy with a thick mat of reindeer lichen covering the ground. Then, approaching the nickel-copper smelters of Zapolyarnyy and Nikel, the scenery changes drastically. Dead tree trunks without any needles left at all. The ground bare and eroded."

Metal smelters in the Russian Arctic, notes AMAP, have created

> industrial deserts, where all or almost all the vegetation is gone. Originally, parts of the Kola Peninsula were covered by bogs, but the mosses disappeared some decades ago from the most heavily polluted areas. Today, an area of 10 to 15 kilometers around the smelters is dry, sandy and stony ground, with only remnants of peat. The Kola Peninsula is one of the eight most seriously polluted "eco-catastrophe" areas of the former Soviet Union.

Concentrations of some heavy metals in the region's freshwater ecosystems have at times exceeded 2,500 times the permissible levels; as a result, "the ecosystems of at least five water bodies are completely destroyed." In an area as far as twenty-five miles from the smelters, concentrations of nickel, copper, cobalt, cadmium, and mercury in the surface sediments of lakes "are 10–380 times background levels."

✳ ✳ ✳

Although some of the worst pollution in the Arctic comes from local sources, much originates far to the south. In the 1950s, pilots in weather observation planes began reporting a strange discolored layer in the sky over the High Arctic, thick enough to obscure visibility and far different from any flying conditions they had encountered before. By the 1980s, this "Arctic haze" had been identified as dry sulfate particles, as well as soot and some soil, from heavily industrialized areas in Europe. It was one of the first indications that the Arctic environment cannot escape the reach of the industrialized world.

In fact, some of the very qualities that make the Arctic so cold and forbidding, and thus contribute to its being sparsely populated and barely developed, may leave it especially vulnerable to pollution from the south. Every winter, an intense high-pressure system forms over Siberia, pushing the cold air down and forcing it to spread out so that the Arctic front moves far to the south, encompassing all of Russia and much of central Asia, covering all of Canada except the warm western coast, and reaching as far as the Great Lakes region of the United States. The front brings bitter winter chills to the area it blankets; it also gathers emissions from European, North American, and East Asian homes, factories, and automobiles into its frigid embrace. Wind patterns are such that air masses from the latter two regions must pass over large stretches of ocean on their way north, during which time low-pressure systems over the water give rise to rain and snow, helping flush the pollution from the atmosphere. Between the European landmass and the Arctic, however, there is no such cleansing process, and the high-pressure systems keep the chances of precipitation to a minimum. At the same time, sunlight, which can break down some contaminants, is largely absent during winter, allowing them to build up until spring, at which time many of the accumulated contaminants return to the ground as dry particles or in rain.

In this way and by other mechanisms, the Arctic is bombarded with a wide array of contaminants from afar: the sulfates and soot that constitute Arctic haze; metals and metal compounds such as those also produced by the smelters at Noril'sk or on the Kola Peninsula; and, of particular concern, a series of human-made, or anthropogenic, compounds collectively known as persistent organic pollutants, or POPs.

POPs include pesticides such as DDT, chlordane, and dieldrin: at one time widely used, such pesticides have been banned in many countries because in addition to killing the insects they were designed to eradicate, they became linked with death and disease among birds, mammals, and humans. It was the severity of their effects on birds that led Rachel Carson to title her groundbreaking 1962 polemic against pesticides *Silent Spring.*

Some POPs, such as dioxins and furans, are by-products of industrial processes that are released into the atmosphere by, for example, waste incinerators, leaded gasoline, and some wood-burning stoves. Still others are chemicals that were praised when first developed for their versatility and that accordingly have been put to use in a suite of applications. Polychlorinated biphenyls (PCBs), probably the most celebrated—or infamous—of them all, have been used as coolants in electrical transformers; as lubricants, hydraulic fluids, and liquid seals; and in weatherproofing, flame retardants, paints, varnishes, and inks. They are, in other words, everywhere. Unfortunately, they are also highly toxic and have been associated with learning disabilities, immune system dysfunction, sperm damage, breast cancer, and miscarriages.*

There are relatively few local sources of POPs within the Arctic itself. Most of them are the result of former—and in limited, mostly illegal instances, current—pesticide use or are related to DEW sites and military bases, where PCBs were used primarily in electrical equipment. A total of around thirty tons of PCBs were used at DEW stations, of which an unknown quantity has ended up in landfills. But the substantial majority of POPs in the Arctic are believed to have made their way north from temperate areas.

Almost all POPs share two principal characteristics that make them of particular relevance to the Arctic. First, as their collective name implies, they are extremely stable; consequently, they can persist in the ecosystem for years— plenty of time for them to be transported, by precipitation, wind, or ocean currents, into the vicinity of the poles. Second, they are lipophilic: literally, they love fat, and accumulate in the fatty tissues of animals—an issue of particular import in the Arctic, where the development of large fat reserves provides essential protection from the cold. These two characteristics allow POPs to bioaccumulate: they build up in one group of animals, and then in the animals that prey on them, and so on up the food chain.

This being so, one would expect the highest levels of POPs to be in top-level predators with large fat reserves. And so it proves. A 1999 study, for example, found PCBs in orcas in Prince William Sound at levels of about 237.7 parts per million (ppm) and DDT at levels as high as 346 ppm. By way of contrast, red meat is considered unsafe for human consumption if it contains PCBs at levels of just 3 ppm, and the limit for DDT in fish is 5 ppm. PCB levels in the Prince William Sound orcas compare to those in belugas in the industrialized St. Lawrence River in Canada, a population of whales renowned for its extraordinarily high rate of lesions, cancers, and other diseases.

---

*Because of such concerns, governments agreed at a conference in December 2000 to an international treaty with the goal of eliminating all POPs.

Polar bears may be at even greater risk—not only because they are top-level predators but also because their preferred prey is marine mammals such as ringed seals, which themselves accumulate high levels of contaminants in their blubber, and because from fall to early spring they undergo one of the most extreme fasts of any mammal, burning up their fat reserves and thus releasing into their systems any accumulated contaminants. Levels of PCBs in polar bears from eastern Greenland, Svalbard, and the McClure Strait are so high that scientists are concerned that the bears' immune systems may be affected. On Svalbard, PCBs have also been linked to an extraordinary development: in recent years, some female polar bears have been found to have male genitalia.

Nor are polar bears and killer whales the only predators at risk. Marine mammals, particularly seals and small whales, play a vital role in the diet of Inuit, Inuvialuit, Eskimos, and other indigenous polar peoples. As a result, despite living thousands of miles from most industrial sources, the bodies of Arctic peoples contain some of the highest levels of contaminants anywhere.

Researchers have found that breast milk in Inuit women contains levels of POPs as much as seven times greater than in women in Europe or the United States and higher than among any other people in the world. A study of an Inuit population at Qikiqtarjuaq, on Baffin Island, found that more than 50 percent were exposed via their diet to levels of POPs that exceeded World Health Organization guidelines; in fact, some ingested more than twenty times the daily limit. Inuit villagers on Broughton Island, which lies just off Baffin Island, to the west of Greenland, have the highest levels of PCB contamination of any human population ever recorded.

The diets of Arctic peoples provide many considerable health benefits. Marine mammals are rich in polyunsaturated fatty acids, which may lower the risk of heart disease; whale skin is rich in selenium, which may guard against both heart disease and cancer; and traditional foods are generally high in protein and many vitamins and minerals and provide a substantial portion of people's energy requirements. Consequently, most researchers have advocated the continuation of those diets because the benefits have generally been considered to outweigh the risks. So far, that still stands: nonetheless, the enormous levels of contaminants Arctic peoples are accumulating is an issue that natives and Western scientists alike are monitoring with concern.

✳ ✳ ✳

In the great ice sheet that covers Antarctica, layer upon layer of snow has captured the telltale signs of the pollution that was in the air before it fell. By drilling through the ice and extracting and analyzing cores, researchers can look back through time: this band shows the era of atmospheric nuclear tests, this one the rise in use of leaded gasoline, and here now the appearance of PCBs and DDT. But although even Antarctica has not remained untouched by the contaminants that bedevil the Arctic, the levels on the great frozen continent and its environs are far lower than those in the North Polar regions. Antarctica is helped by its isolation, and in contrast to the weather patterns that invite passage of pollutants into the Arctic's domain, the swirling belt of winds and currents surrounding the Antarctic Continent helps to ward off the unwelcome attentions of contaminants from the industrialized world.

But it is not a wholly effective barrier; if it were, there would be no sign of twentieth-century pollution—except that imported by tourists, researchers, and explorers   at all. And indeed, in the same way the low-pressure systems that build up over Siberia expose the Arctic to the effects of contaminants from the south, that great band that otherwise isolates Antarctica left it surprisingly and uniquely vulnerable to an environmental threat that emerged suddenly and unexpectedly in the 1980s.

The Hole in the Sky

Ozone is a pale blue gas that is poisonous to humans and yet is a prerequisite for life on Earth. At or near ground level, it is noxious and sometimes pungent and is often a major component of smog. But in the stratosphere—the layer of atmosphere from about ten miles to around thirty miles above Earth's surface—it is a godsend, without which life as we know it could not exist.

Stratospheric ozone is formed when the sun's ultraviolet radiation splits apart molecules of oxygen. The chemical symbol for oxygen gas is O_2, signifying that it is composed of two oxygen atoms; under the action of the sun's rays, oxygen breaks down, creating free oxygen atoms, which swiftly become attracted to existing molecules of oxygen gas to form ozone, O_3. Ozone is then itself split asunder by ultraviolet rays of a different wavelength, but its molecules swiftly reform. In the course of this rupturing and rebuilding, ozone acts as a defensive shield, absorbing much of the sun's ultraviolet radiation before it reaches the ground.

That is important because although the sun's rays bathe the planet in warmth and energy, some of those at the ultraviolet end of the spectrum—particularly those known as ultraviolet-B, or UV-B, rays—can be dangerous and even fatal to some forms of life. It is the sun's UV-B rays, for example, that give beachgoers suntans but that in too great a quantity result in sunburn. Prolonged exposure can lead to certain forms of cancer, including the frequently fatal malignant melanoma. It was that fact which caught people's attention when scientists Sherwood Rowland and Mario Molina announced in 1974 that Earth's ozone layer—and hence the protective blanket it draped over the planet—was in imminent danger of disappearing.

The cause of the scientists' alarm was the group of anthropogenic chemical compounds known as chlorofluorocarbons, or CFCs: widely used primarily as propellants in aerosol sprays but also, among other things, as coolants in refrigerators and air conditioners. Other researchers had already noticed that CFCs are extremely stable, very hard to break down, and happy to linger in the atmosphere for long periods of time. In fact, Rowland and Molina calculated, they are so stable that instead of ultimately being broken down by sunlight or flushed from the atmosphere in rain or as dry particles, they stay in the atmosphere long enough to make their way slowly up to the stratosphere. There, finally, they meet their match in the form of ultraviolet radiation from the sun. The sun's rays split CFC molecules apart, releasing chlorine atoms. And that, Rowland and Molina calculated, was where the trouble started. According to their equations, those chlorine atoms would collide with ozone molecules, breaking down the ozone by stripping away the third oxygen atom to form chlorine monoxide. Worse, chlorine monoxide is itself an unstable compound and breaks down easily, leaving the chlorine atom free once more; in this way, a single chlorine atom could destroy several thousand molecules of ozone.

Because CFCs were relatively new and because of the time it took them to

Chapter 7. The Ends of the Earth 219

reach the stratosphere, it would, Rowland and Molina predicted, be some time before their effect would be noticed. But because they are so persistent, even if all production of CFCs were to cease immediately, those that had already been released into the atmosphere would continue to wreak havoc for years, even decades, to come. And if production continued at present rates, within 100 years there could be enough CFCs in the atmosphere to remove as much as 13 percent of the global ozone layer, more than enough to have a serious effect on life on Earth.

Rowland and Molina published their findings in the scientific journal *Nature;* within a year, two other groups of scientists published similar studies. At first, the response was swift: Congress held hearings, and by 1978 the United States had imposed a unilateral ban on the use of CFCs as propellants in aerosol spray cans. But CFC uses in other U.S. industries, not only in refrigerators and air conditioners but also, for example, in the Styrofoam boxes used to carry home Happy Meal dinners, were not regulated and were in fact increasing in quantity, as was the use of CFCs, in aerosols and elsewhere, in other countries. After a brief period of decline following the U.S. aerosol ban, emissions of CFCs into the atmosphere began increasing again in the early 1980s.

Researchers such as Rowland and Molina insisted that unless CFC production and emissions were severely curtailed or eliminated—in all industries and all countries—it would be only a matter of time before the ozone layer showed signs of serious depletion. Nobody, however, expected the first sign to be quite so huge. And no one expected it to appear over Antarctica.

* * *

In 1982, scientists working at the British Antarctic Survey (BAS) base at Halley Bay, on the southeastern coast of the Weddell Sea, made a startling discovery. Ever since the International Geophysical Year, researchers at Halley had been measuring ozone levels in the Antarctic atmosphere; during all that time, even after the controversy erupted over the ozone layer and CFCs, their findings had shown no change of any great moment. Now, all of a sudden, their measurements were showing a dramatic drop in ozone levels above Antarctica: not just 1 to 4 percent—the kind of figures that had been bandied about during debates over a CFC ban—but more than 20 percent. It was so incredible, so off the predicted scale, that the immediate reaction of Joe Farman, the scientist heading up the research, was that something had to be wrong with the equipment. The Halley station used ground-based instruments called Dobson photospectrometers, and although Farman and colleagues had not previously had cause to doubt their reliability, the instruments were old and could sometimes be temperamental. Besides, with or without CFCs, ozone levels tend to fluctuate naturally in accordance with chemical processes in the atmosphere and the activity of the sun. Moreover, a U.S. weather satellite, *Nimbus-7,* which had been launched in 1978 and carried two instruments to measure ozone levels, had not returned data showing anything as remarkable as what the Halley team was seeing.

So the BAS researchers sat on their findings for a year. They even kept quiet the following year when their readings showed yet greater declines in ozone levels. Then, in 1984, they used a newer photospectrometer. They also set up a second measuring station on Argentine Island, 1,000 miles to the northwest. This time, there was no doubt. If anything, ozone losses were increasing: according to BAS, there was a 30 percent drop during October 1984 alone.

Farman and colleagues went public and published their findings in *Nature.* Reaction ranged from shock to disbelief. Few of the researchers working on ozone–CFC issues had heard of Joe Farman, or even of BAS; who were these people, they asked, who were now claiming such hugely improbable ozone losses? Besides, if the situation was that dire, why had *Nimbus* 7 not detected it?

In fact, it turned out, the satellite *had* registered the losses; it was just that the levels involved were so improbable that the computer analysis program that interpreted the data had been instructed to all but ignore them. Ozone levels are measured in what are called Dobson units; prior to widespread production and emission of CFCs, ozone layer measurements would sometimes produce readings of 300 Dobson units or more, but BAS figures showed that above Antarctica, that had dropped by nearly half, to almost 150 Dobson units. Previously, there had been next to no readings of less than 200 units, and the National Aeronautics and Space Administration (NASA) programmers responsible for *Nimbus* 7 had instructed their computers to treat any measurement less than 180 units as being in error. The program did note these low readings and flag them as anomalies to be checked later, but because of the amount of work involved in processing data from a whole range of satellites, researchers simply had not had time to go through all the readings. When Farman and his team published their results, however, the *Nimbus* scientists instructed their computers to reanalyze the data, this time no longer assuming that extremely low measurements were mistaken. The result was one of the defining images of the environmental movement of the late twentieth century: a computer-enhanced image showing extremely low ozone levels over Antarctica—what was swiftly dubbed an ozone "hole"—over much of the continent.

Subsequent studies at McMurdo Station, using not only photospectrometers but also instruments attached to weather balloons, refined the information and provided new data: the damage to the stratosphere was confined to a layer between 7.5 and 12.5 miles in altitude; on average, overall depletion within that band was about 35 percent. But between 8 and 11 miles, the stratosphere had lost more than 70 percent of its ozone, and at one point in October 1986, 90 percent of the ozone had disappeared from a layer between 1 and 3 miles thick.

The "hole" emerged each austral spring and steadily repaired itself through summer but then reappeared with the demise of winter and the dawn of spring. Each time, a part of the ozone layer bigger in area than the United States and deeper than Mount Everest disappeared from the stratosphere.

An ozone hole over Antarctica was clearly not going to result in a rise in human cancers—although the discovery of lower levels of ozone depletion farther north, over Australia, New Zealand, and southern South America, prompted very real and ongoing human health concerns. But there was genuine apprehension that increased UV-B levels could have significant adverse effects on the ecosystem of the Southern Ocean. And indeed, researchers found that UV radiation could and did cause some algae to photosynthesize less or not at all and even resulted in reductions in entire phytoplankton communities. But in general, the effects have been less severe than predicted worst-case scenarios: some phytoplankton species have proven more adept than expected at protecting themselves from UV radiation's harmful effects, and such effects are apparently to some extent mitigated by the fact that the ozone hole's peak occurs at a time when sea ice, which absorbs most of the UV radiation, is still extensive. But even if phytoplankton are relatively well protected from the ozone hole's consequences, the same may not be true for krill: a 1999 study showed that even relatively low levels of UV radiation can cause DNA damage and increased mortality, especially in juveniles. Given krill's vital role in the Southern Ocean food web, the repercussions could prove to be significant.

As scientists tried to determine the effects of decreasing ozone levels, governments responded to public concern and acted to stop the problem from getting worse. For environmentalists, the 1987 Montreal Protocol on Substances That Deplete the Ozone Layer was both a tremendous success—in that it committed signatories to end production and use of CFCs and other ozone-destroying chemicals, such as halon, methyl bromide, and hydrochlorofluorocarbons (HCFCs)—and a deep disappointment, in that such bans would not be immediate or even imminent. Industrialized countries agreed to phase out CFCs and most other ozone-depleting substances by 1996 (except methyl bromide, which does not have to be phased out completely until 2010, and HCFCs, production and use of which are being ratcheted down but do not have to reach zero until 2030). Less industrialized countries pleaded for more time and were granted it: they have until 2010 or 2015 to phase out most substances and until 2040 to eliminate HCFCs.

The Montreal Protocol at first led to a decline in the amount of ozone-destroying chemicals entering the atmosphere; since then, however, as some industrialized countries have balked at meeting their commitments and with the reductions by less developed countries yet to kick in, that decline has shown signs of bottoming out. It will, presumably, at some stage resume its downward trajectory. However, even if all production and use of CFCs and associated chemicals were to stop tomorrow, the compounds' persistence and the time it takes them to travel up through the troposphere and into the stratosphere mean that it will probably be several decades before the ozone layer starts showing any significant recovery. In 1995, the area of severely depleted ozone above Antarctica was as large in area as all of Europe and twice as large as in the two preceding years, and by 1998 it was larger still. After a decline in 1999, it expanded once more in the spring of 2000 to an area of 11 million square miles—three times larger than the land area of the entire United States and the largest yet recorded.

Although in hindsight the Antarctic ozone hole is often seen as the driving force behind the Montreal Protocol, negotiations—spurred by the warnings of Rowland, Molina, and other scientists—were already under way when Joe Farman and colleagues published their paper. Increased public concern brought about by discovery of the ozone hole doubtless provided that final political momentum, but at the time, says Richard Elliot Benedick, the chief U.S. negotiator, the hole was considered an anomaly. In fact, he said, "disturbing new evidence [that the hole was caused by CFCs] was not, in actuality, completely analyzed and released until six months after the treaty was signed."

Not only was the ozone hole a surprise, one that had confounded researchers and gone unpredicted by computer models, but for some time its cause remained a mystery. To some, its sheer improbability suggested it must be a natural phenomenon, perhaps related to phases in solar activity and interactions with natural climatic conditions rather than anything to do with CFCs. After all, CFCs were predominantly manufactured and used in the north—in the United States, in Europe. Why would their effect first be seen above Antarctica?

The answer is that in much the same way conditions in the Arctic encourage the region's contamination by pollutants from farther south, the profound cold above Antarctica creates the perfect conditions for CFCs to attack the ozone layer. As we saw earlier, during the long Antarctic winter night, circumpolar winds that race around the continent all but seal it off, creating a so-called polar vortex in which cold air in the stratosphere drops to temperatures around −120°F. Such extreme cold allows the formation of thin, wispy polar stratospheric clouds, the icy surfaces of which provide a platform for chemical reactions in which chlorine is freed from compounds such as CFCs. Over the course of winter, levels of free chlorine continue to build up, and on the return of spring, sunlight prompts the process that causes chlorine to attack ozone.

The discovery that this was the mechanism involved in ozone-hole formation prompted another realization: given that similar dynamics operate in the Arctic atmosphere, albeit on a smaller, warmer, scale, perhaps some form of ozone hole could develop there, too. In 1988, an international expedition headed by NASA and the National Oceanic and Atmospheric Administration (NOAA) found concentrations of chlorine compounds in the Arctic atmosphere that were fifty times higher than normal; in February 1989, a NOAA researcher concluded that the Arctic atmosphere was "primed for ozone destruction." But the Arctic vortex is less stable than that which swirls around Antarctica, and stratospheric temperatures are warmer, and no ozone hole had developed at that time. Ozone levels did decline, however, and then, in 1997, ozone levels over the Arctic showed a catastrophic drop as had those above Antarctica more than a decade earlier. Recorded levels were 40 percent below the average for 1979–1992, and since then they have dropped yet more. In spring 2000, levels were at their lowest yet, and all the signs are that the Arctic ozone hole is likely to continue growing for some time—not only because of the lag effect mentioned earlier but also because conditions are becoming more favorable for its growth. The stratosphere above the Arctic is growing colder and colder—still warmer than that above Antarctica but increasingly suitable for the formation of polar stratospheric clouds and subsequent initiation of the process of ozone depletion. And the reason for this stratospheric cooling stems from another, seemingly contradictory problem: the atmosphere below it is growing warmer.

The Global Greenhouse

The process works something like this. Energy from the sun warms the surface of Earth; in turn, Earth radiates some of that energy, at longer wavelengths, back toward space. Gases in the lower atmosphere—principally carbon dioxide and water vapor but also, to a lesser extent, methane and nitrous oxide—absorb a percentage of that energy, radiate some of it back toward the planet's surface, and again absorb some of the energy that the surface reflects back upward. Although the analogy is not entirely accurate, this has been termed the "greenhouse effect," in recognition of the fact that a greenhouse's glass allows the sun's radiation to pass through unimpeded but absorbs some of the radiation that is re-emitted by the plants and soil, making it warmer inside than outside.

A little greenhouse effect is a wonderful thing: through a natural series of feedback loops involving atmosphere, ocean, land, and even life itself, it keeps Earth comfortably warm. But rapid alteration of any of the elements involved—for example, increases in one or more of the greenhouse gases—

could upset the system. The lower atmosphere might absorb more of the energy radiated from Earth's surface, causing less energy to reach the stratosphere and resulting in a stratospheric cooling, such as that now seen above the Arctic. Of at least equal significance, however, it might, by trapping more heat, lead to considerable increases in Earth's surface temperatures.

This is not a new idea. Back in 1896, Swedish scientist Svante Arrhenius calculated that if all else remained the same and levels of atmospheric carbon dioxide doubled, the net result would be an increase in the average global temperature of around 5–6°C, or 9–11°F. Although that is about twice the increase that, until recently, most modern calculations assumed would be the case, the principle remains sound; furthermore, it is testable. Even as Arrhenius worked on his figures, levels of carbon dioxide in the atmosphere were increasing, and they are continuing to do so more than a century later.

The principal cause of this increase is the burning of fossil fuels, such as coal and oil. Because these fuels contain carbon, their combustion returns carbon to the atmosphere, where it reacts with oxygen to create carbon dioxide. There are other factors at play—tropical forest fires, for example, also release huge amounts of carbon, and flatulent livestock has been fingered as a major source of atmospheric methane—but fossil fuel combustion, beginning around 1750 with the start of the Industrial Revolution and spiking upward since the middle of the twentieth century, is by far the biggest contributor. Measurements of atmospheric particles deposited over millennia in the ice sheets of Greenland and Antarctica have shown that for several thousand years, levels of atmospheric carbon dioxide were generally around 280 ppm; in the mid-1950s, a series of readings suggested that levels had reached around 315 ppm, and since then they have soared above 360 ppm—the highest level in more than 400,000 years.

In December 1997, responding to growing alarm over this trend, governments concluded an international agreement, the Kyoto Protocol, under which the countries of the European Union, for example, agreed that by 2008–2012 they would reduce their emissions by an average of 8 percent from 1990 levels. The United States agreed to a reduction of 7 percent. But few of the major industrialized countries have shown any inclination toward ratifying the treaty, and the U.S. Congress remains adamantly hostile. Congressional objections stem from an insistence on the part of the fossil fuel industries that increased levels of atmospheric carbon dioxide are having little or no discernible detrimental effect on global weather patterns; indeed, a number of researchers argue that any such changes are the result of the myriad natural processes that influence Earth's climate. Even so, the evidence increasingly suggests that such increases in atmospheric carbon dioxide are having a measurable effect on global temperature.

According to the World Meteorological Organization (WMO), an arm of the United Nations, the twentieth century was the warmest century of at least the past millennium, and the 1990s were the warmest decade of that century. Seven of the ten warmest years on record were in the 1990s—with the five warmest, in descending order, being 1998, 1997, 1995, 1990, and 1999—and the other three were all in the 1980s.

The warming is not uniform. In England, where records stretch back to the late 1630s, 1999 was the warmest year on record; in Japan, it was the third warmest year in 102 years of records; in Canada, the second warmest since 1948. Indeed, the vagaries of climatic systems are such that some areas—much of the southeastern United States, for example—are actually experiencing a slight cooling. Equatorial regions are, in general, so far showing little temperature increase. The largest increases, and the areas where the most dramatic evidence of global warming can be found, are at the ends of the Earth, in the Arctic and Antarctic.

* * *

The Prince Gustav Channel separates the mainland of the Antarctic Peninsula from James Ross Island and Vega Island, to the east. It is a small channel, five to fifteen miles wide and eighty miles long; for much of its existence, passage through it was blocked by a permanent ice shelf. Otto Nordenskjöld traversed this ice shelf during his enforced sojourn in the region a century ago, and in so doing he completed the first circumnavigation of James Ross Island.

Today, the only way to circumnavigate James Ross Island is by ship. In February 1995, ninety-two years after Nordenskjöld sledded across the ice around the island, the ice shelf splintered and broke apart. Alerted by satellite observations, British Antarctic Survey observers flew over the scene and reported "a dense plume of fragments extending several hundred kilometers into the sea." In an instant, the Larsen-A Ice Shelf—an area of ice the size of Luxembourg— had gone, and the geography of Antarctica had changed forever. "In 25 years of Antarctic field work," said BAS geologist Mike Thomson, "I have never seen anything like it."

And where Larsen-A went, its larger cousin immediately to the south, Larsen-B, seems set to follow: at the same time Larsen-A disintegrated, Larsen-B calved a giant iceberg, roughly 49 miles by 23 miles, and now the rest of the shelf is also showing signs of imminent collapse. Large cracks and fissures crisscross its surface; in just a few months between late 1998 and early 1999, more than 650 square miles broke away. For the nearby Wilkins Ice Shelf, the prognosis is, if anything, even more alarming. In March 1998 alone, it lost 425 square miles and retreated about 22 miles. Both ice shelves had shown signs of retreating since measurements began, around the time of Operation Tabarin in the 1940s, but over the course of five decades or so, Larsen-B and Wilkins had, combined, lost about 2,700 square miles. To suddenly lose nearly 1,200 square miles in the course of just one year was, as David Vaughan of BAS succinctly put it, "clearly an escalation." Within a few years, he noted, most of the Wilkins Ice Shelf would very likely be gone; Larsen-B may not be far behind.

The search for a culprit to explain such losses swiftly yields an obvious suspect. Whereas global average temperatures have climbed over the past few decades, those on the Antarctic Peninsula have soared: year-round by 3–4°F, and during winter by a remarkable 7–9°F.

Such rising temperatures have resulted in more than disappearing ice shelves: southern elephant seals and fur seals and gentoo and chinstrap penguins, species that normally breed farther north, are establishing colonies at increasingly southerly latitudes. Shrubs and mosses never before seen in the region are becoming established. Meanwhile, other species in the region are declining: the Adélie penguin population around the U.S. research site Palmer Station has dropped by half—from around 15,000 pairs to a little more than 7,500—in twenty-five years and by 10 percent in just two years. According to Bill Fraser of Montana State University, the probable cause of the Adélies' decline is disappearing sea ice: unlike chinstraps, Adélie penguins rely heavily on winter sea ice, gathering en masse and feasting on young krill that gather around its edges. Less sea ice may mean both fewer young krill and less area on which the penguins can gather, either or both of which could explain the Adélies' rapid disappearance.

* * *

As global temperatures rise, so do global sea levels: by 4 to 10 inches over the twentieth century, according to the Intergovernmental Panel on Climate Change (IPCC), a body of scientists established under the auspices of the United Nations. By 2100, under what is termed the "business-as-usual" scenario—in which greenhouse gas emissions continue at present-day levels— that number could increase by an additional 19 inches. And although some estimates suggest that the total rise over the next century could be as little as 6 inches, others postulate it could be more than 36 inches. For low-lying coastal areas as in Bangladesh or the Nile River delta, or for island states such as the Republic of Maldives, in the Indian Ocean, that much of an increase in sea level could prove devastating.

The bulk of such sea-level rise is caused by the fact that as water heats up, it expands; most of the rest is caused by the calving and retreat of mountain glaciers, including those in the Arctic. But the contributions of such processes would be as nothing compared with what would happen under the nightmare scenario in which global temperatures increase so much that they cause the melting of the Antarctic ice cap.

The collapse of the Larsen and Wilkins Ice Shelves—or of the Wordie Ice Shelf, which has been steadily disappearing over a longer period—will not in itself contribute to sea-level rise. These shelves are already floating, so their disintegration does not add to the total amount of water in the ocean, any more than melting ice cubes cause a drink to overflow from a glass. If, on the other hand, warming increased enough to break down the Ross and, especially, Ronne-Filchner Ice Shelves, then all bets would be off. Those shelves essentially act as "dams" for the ice streams that flow from the heart of Antarctica; one disquieting prediction is that should they disappear, those streams might begin disgorging into the warming waters around the continent, leading to diminishment and, perhaps, sudden collapse of the West Antarctic Ice Sheet. Unlike the larger ice sheet that covers East Antarctica and is mostly grounded by bedrock, the foundation of the one that covers West Antarctica is below sea level and thus is already vulnerable to the nibbling and melting effects of warming waters. Were the West Antarctic Ice Sheet to collapse, even the most pessimistic sea-level rise scenarios would be washed away and the ocean would rise by an estimated 13–20 feet. The result would be catastrophic: in the words of one scientist, the "oceans would flood all existing port facilities and other low-lying coastal structures, extensive sections of heavily farmed and densely populated river deltas of the world, major portions of the states of Florida and Louisiana, and large areas of many of the world's major cities."

Fortunately, so far that eventuality seems a long way off, if indeed it is to be expected at all. Both the Ross and Ronne-Filchner Ice Shelves have average summer temperatures of 12°F; a full twenty degrees of warming would be needed to push them to the melting point, and that is outside any of the models predicted by the IPCC. In fact, there is no sign of any warming in Antarctica outside the Antarctic Peninsula region, and parts of the interior may even have cooled. If average temperatures were to rise over Antarctica as a whole, one of the early consequences might actually be a net gain in the size of the ice cap as warmer conditions initially lead to increased precipitation, causing more snow to fall. Most researchers believe that if the ice sheet were eventually to collapse, it would probably be over a relatively long time period, in the range of five to seven centuries.

Yet even if the ice sheet itself does not disintegrate, the melting of ice shelves could have serious and wide-reaching implications. Although the precise mechanism involved is poorly understood, ice shelves are known to be integral to the process by which cold surface water sinks to the ocean bottom and then slowly spreads northward, north of the Antarctic Convergence, and into all the world's seas. If the ice shelves were not there, it has been suggested, that water could never sink, and the entire global "conveyor belt" of ocean currents that regulates Earth's climate would be altered or would even grind to a halt. So far, those shelves that have collapsed and those that are rapidly disintegrating are too small to have any significant effect on this thermohaline conveyor. The real danger would come with changes in the Ronne-Filchner Ice Shelf, and that, as we have seen, would seem to be a way off just yet.

But changes in the ice shelves of Antarctica are not alone in their potential to influence the conveyor belt. Some researchers argue that a very similar process could happen in the Arctic, and that in the process a global warming could, paradoxically, cause a European cooling. Moreover, there is a precedent.

* * *

Twelve thousand years ago, Earth was slowly warming, recovering from a long ice age that had lasted the better part of a hundred thousand years and had begun to release its frigid hold just a few millennia previously. Forests that had retrenched during the ice age began expanding once more. A climate that had been predominantly cool and dry grew ever warmer and more moist. Humans took advantage of the more hospitable conditions, hunting game, gathering seeds and berries, and perhaps becoming established in large, permanent settlements.

But then, abruptly, before it fully had a chance to flourish, this Eden was forced into retreat and the world was once more plunged rapidly into a glacial freeze. According to the preferred explanation, rising temperatures had melted the giant Laurentide Ice Sheet, which swathed much of eastern North America, causing a massive influx of freshwater into the ocean. Cold ocean water south of Greenland was no longer dense enough to sink to the bottom—a process that, as in the Antarctic, is essential to kick-start the global conveyor belt—and as a result the ocean currents that had brought warmth to the North Atlantic Ocean simply shut down. Cold conditions persisted for about 1,300 years, and then suddenly—very suddenly, in fact, over a span of just a few decades—the system apparently corrected itself and the climate returned to its previous warm, interglacial path.

The Laurentide Ice Sheet is long gone, and nothing comparable in size remains in the Northern Hemisphere. The nearest equivalent is the ice cover over Greenland, and data are inconclusive as to whether or not that ice sheet is melting. One study has suggested that the southern part of the ice sheet is losing about two cubic miles per year while precipitation over the central plateau is such that overall, the ice sheet is actually gaining in extent. The report of a NASA study in July 2000 argued that the center is neither gaining nor losing snow and ice cover, but the edges of the ice sheet are definitely melting enough to contribute to sea-level rise—albeit by a tiny amount thus far, just five-thousands of an inch annually.

Whatever the status of the Greenland ice sheet, an interruption in the ocean conveyor is still within the realm of possibility: rising sea surface temperatures or increased precipitation may also warm or freshen the ocean sufficiently to tip the system over the edge. According to some researchers, a doubling of atmospheric greenhouse gases from preindustrial levels—which some scenarios predict could happen by the end of the twenty-first century if remedial measures are not undertaken—would be enough to shut the conveyer down. Whatever the likelihood of such a scenario, scientists recognize that were the conveyor belt to grind to a halt, there would be little or no warning. There would be no gradual change in conditions: suddenly, almost overnight, it would just stop.

✳ ✳ ✳

For close to three decades, George Divoky, a scientist with the University of Alaska Fairbanks and the Alaska SeaLife Center, has set up camp almost every year on Cooper Island, just off the North Slope, near Barrow. At times, it has been an uncomfortable experience. He recounts, for example, the time he heard something outside in the middle of the night and emerged the next morning to find polar bear tracks leading right up to the edge of his tent, just inches from his head. But Divoky has stuck it out, returning year after year to conduct field research. In the process, he has gathered evidence of what he says is "among the first documented biological effects of climate change in the Arctic."

The canary in Divoky's global warming coal mine is a species of seabird called the black guillemot. Found throughout the higher latitudes of the Northern Hemisphere—in terms of the number of colonies, black guillemots are the most abundant seabird in Britain—they have not been common in northern Alaska since probably at least the early seventeenth century, when global temperatures began to fall and snow cover increased. The guillemots take eighty days between occupying a nesting site on or near the ground and fledging their young; however, says Divoky, for the best part of four hundred years, there were relatively few years in which northern Alaska experienced eighty consecutive snow-free days.

As recently as 1966, only one nesting site was known on Alaska's northern coast. But in 1972, Divoky discovered a small colony of ten pairs on Cooper Island, nesting in the remnants of ordnance that had been blown up on the island in the 1950s. Then, from 1975 to 1990, the number of birds in the colony soared, peaking at 225 nesting pairs. This increase was fueled, says Divoky, by warmer temperatures causing earlier snowmelt in the spring, providing greater access to nesting sites and a longer window during which the birds could breed.

Beginning in 1990, however, numbers began dropping rapidly; presently, the Cooper Island colony is down to 110 pairs and falling, and the birds that remain are lighter and thinner. Mortality among birds that nest on Cooper Island is higher, and there are fewer immigrant birds from Wrangel Island, in the Russian Arctic, which Divoky believes is the source colony. The reason, says Divoky, is again climate change, this time because of its effects on sea-ice cover. Black guillemots use sea ice to rest and feed on the arctic cod that live beneath floes. But Arctic sea ice is thinning and retreating, providing less refuge, reducing the available habitat for arctic cod, and forcing the birds to fly farther in search of food.

There is growing support for Divoky's contention. Three studies published at the end of 1999 presented evidence that, the authors claimed, showed significant declines in Arctic sea ice. One, based on sonar data from nuclear submarines, argued that the perennial ice cover of the Arctic Ocean is about 40 percent thinner than it was two to four decades ago; a second, using satellite imagery, concluded that the ice had also shrunk in extent, by 14 percent from 1978 to 1998. A third combined forty-six years' worth of data from five different sources, including ground-based observations and satellite data, and found that total sea-ice extent—including seasonal as well as permanent sea ice—has since 1978 been shrinking by about 143,000 square miles per decade. An additional study published in 2000 predicted that if current trends continue, summer sea ice will disappear entirely from the Arctic beginning around 2050.

Few, even global warming skeptics, dispute that Arctic sea ice has been decreasing over recent decades: the raw data are there for all to see. The question is whether the fifty or so years over which measurements have been made is a long enough span to show any actual trend, and even if the long-term direction is in fact downward, whether it is the result of anthropogenic climate change or simply part of a natural cycle. The authors of the third study mentioned earlier compared their results with computer models and found that there was less than a 2 percent chance that the melting was due to normal climatic variation.

Not everyone is completely sold on the models' accuracy, and even some of those who are generally sympathetic to the notion that climate change is affecting the Arctic are not entirely convinced; other researchers counter that the models are among the most sophisticated ever developed. Either way, the potential consequences of retreating sea ice are profound indeed. The Arctic's ice cover helps maintain the cold conditions that created it; it reflects solar radiation and insulates the relatively warm ocean below from the cold Arctic air. If the ice disappears, the ocean will retain more heat from the sun—as much as 100 times more than does the equivalent area of sea ice—and at the same time, a greater proportion of that energy will be released into the atmosphere, further warming the climate and leading to more melting, and so on. In addition, a transition from ice-covered ocean to open water may affect ocean circulation in ways that are not yet fully understood.

Then, of course, there are the effects on wildlife—and George Divoky's black guillemot is not the only species that could be affected. Like the seabirds, many marine mammal species, such as ringed seals and walruses, use sea-ice cover to rest or to feed on the arctic cod that shelter beneath the floes. The distribution of bowhead whales, belugas, and narwhals is also strongly tied to the availability of prey, such as fish or plankton, along the ice edge. Sea ice is integral to the life of polar bears as well: although they spend some time onshore, especially to breed, and can swim perfectly well in the water, sea ice is the polar bears' domain. Indeed, in many languages, "ice bear" is the polar bear's name. It uses the ice to cover vast distances and to hunt for seals, its preferred prey. Without sea ice, the polar bear would essentially have no habitat and no easy way to catch prey. Although there is no evidence that the species is declining, there is already one region where some researchers believe the first signs can be seen of what the bears have in store for them in a warming world.

Hudson Bay, in the east of Canada and at the very southernmost reaches of what could be considered the Arctic Ocean, is home to one of the most extensively studied of all polar bear populations. A 1999 study showed that the bears in this population are losing weight and having fewer cubs—a trend that appears to be correlated with progressively shorter ice seasons in western Hudson Bay. In particular, spring breakup, which once came in July, now comes about three weeks earlier—a matter of some seriousness for polar bears, given that they must fast for long periods, as much as eight months in the case of pregnant females, when the ice is gone. Early breakup means not only a longer fast but also less time to gorge on seals and build up fat reserves for the lean months ahead. Although the Hudson Bay polar bears are not yet in decline, the study's authors note that if these trends continue, the population will eventually be unable to sustain itself.

* * *

In the bleak High Arctic environment of Bathurst Island, in recent years the ground has at times been littered with the carcasses of hundreds of Peary caribou. The Peary, a smaller subspecies of the caribou and reindeer that roam lower latitudes, inhabits the islands of the Canadian High Arctic, an area that appears to merit the overwhelming image the region conveys, of a harsh, frozen, seemingly almost lifeless realm. Peary caribou, however, not only have survived in this forbidding landscape—eking out a living on patches of lichen, moss, grass, and small shrubs just a few hundred miles from the North Pole— but have thrived. Although the Peary caribou have probably never been especially numerous, there were, according to one estimate, almost 24,500 animals in the region in 1961. Subsequently, however, the numbers have plunged. A 1997 survey yielded an estimate of just 1,110 caribou. Around Bathurst Island, where 2,400 animals were counted in 1993, only 43 were found five years later.

As temperatures have increased over the western Arctic, low-pressure systems have brought warm, moist air masses from the south, resulting in the normally dry High Arctic receiving unprecedented levels of snow. Layers of alternately freezing and melting snow have blanketed the ground where the Peary caribou forage, erecting a barrier between the animals and their food and causing them to starve to death.

The fact that black guillemots on Cooper Island should benefit from reduced snow cover and Peary caribou on Bathurst Island should suffer from increased snowfall is demonstrative of the fact that even within a specific geographic region such as the Arctic, a changing climate can take many different directions—and accordingly result in significantly varied consequences for the region's wildlife. While Peary caribou decline in the Canadian High Arctic, many caribou populations at lower latitudes are increasing in number. In the eastern Canadian Arctic, they may actually be benefiting from an anomalous cooling centered around Baffin Island; conversely, it has been suggested that in eastern Alaska and western Canada, longer, warmer summers may prove beneficial to the Porcupine caribou herd.

In the same way urbanization has hindered specialist species and allowed opportunists such as gulls, crows, and foxes to thrive, so will there be winners and losers in a world affected by global warming. For a devastating illustration of the spread of one such global warming beneficiary, drive on the Seward Highway from Anchorage, along the Kenai Peninsula, to Homer. Large stretches of the highway are flanked not only by imposing mountains or the crystal clarity of Cook Inlet but also by swaths of conifers, the boreal forest that by many measures stands as the defining ecosystem of the subarctic. But it takes little time to realize that all is not well with the trees: huge stretches of them are dry and colorless, weak, dying, or dead. The culprits are two insect species: the spruce bark beetle and the western black-headed budworm.

After hatching, the budworm burrows its way into the buds of Sitka spruce and uses silk to tie them shut. After feeding on the buds through spring, if conditions are favorable, the larvae emerge en masse as caterpillars in summer, swarming over the trees and eating the needles. Although native to Alaska, the budworm has recently been particularly virulent, causing reduced growth and increased mortality in trees over an area of 100,000 acres from 1993 to 1998.

The damage from the budworm, however, is as nothing compared with that wreaked by the spruce bark beetle, which has killed Sitka–white spruce hybrids over an area in excess of 3.2 million acres. The bark beetle uses sensitive antennae to detect subtle chemical signals from spruce trees, preferring to lay its eggs in trees that are relatively healthy but slightly stressed or weakened. After finding a tree and laying its eggs, it emits hormones that attract more beetles. The tree calls on its natural defenses to try to repel the invaders but eventually dies when the sheer number of insects overwhelms it.

According to several researchers, including Glenn Juday of the University of Alaska Fairbanks, the insects' increased destructiveness is directly related to a warming climate. Recent warm summers have enabled the bark beetle, for example, to halve the length of its life cycle from two years to one, effectively enabling its populations to double in size. At the same time, warmer, drier conditions are stressing and weakening trees, making them more vulnerable to insect attack. Juday says that the question "is not how much the forests will de damaged. This entire forest system is dying, and the question is what will replace it." According to some researchers, this is a pattern likely to be repeated around the Arctic's fringes. As much as half, say some, of the world's boreal forests could disappear within decades.

✳ ✳ ✳

In 1995, the IPCC declared that the "balance of evidence suggests a discernible human influence on global climate" and predicted a temperature increase of anywhere from 1.8°F to 6.3°F by the end of the twenty-first century. Five years later, as this book was being edited, a draft summary of a new IPCC assessment was distributed to policy makers; its language was more blunt, and its conclusions more confident, than had been the case half a decade earlier. It noted that there was now a "longer and more closely-scrutinized temperature record" and that new computer models allowed for more accurate and sophisticated analysis and prediction. There was now, the assessment stated, "stronger evidence for a human influence on global climate"; furthermore, it estimated that Earth's average surface temperatures could be expected to increase by 2.7°F to nearly 11°F—substantially higher than earlier IPCC estimates and close to the figures first postulated by Svante Arrhenius in 1896.

Despite the significantly larger upper estimate of increase in average surface temperature, a major loss of ice in the West Antarctic Ice Sheet remains "very unlikely," according to the assessment, at least during the twenty-first century. Northern Hemisphere snow and ice cover, however, is expected to decrease further. A local annual average warming of about 5.4°F, sustained for millennia, would be enough to almost completely melt the Greenland ice sheet. A warming of around 10°F would result in a contribution from Greenland to a sea-level rise of around ten feet over a thousand years.

As we have already seen, some researchers continue to cast doubt on the phenomenon of global warming, arguing that any such changes are entirely the result of natural processes, and certainly it would be wrong to understate both the importance of long-term climatic fluctuations and the fact that our knowledge of such fluctuations is incomplete. But atmospheric carbon dioxide levels are at their highest in at least 420,000 years, and the rate of increase is unprecedented in at least the past 20,000 years; the IPCC is now confident enough to express the view that "the observed warming over the last 100 years is unlikely to be entirely natural in origin."

Whatever the causes, it is clear that Earth is undergoing a period of profound climate change. At the center of that change are the polar and subpolar regions, which are experiencing change more swiftly and more severely than anywhere else on Earth, with potential consequences for the environment, wildlife, and human inhabitants both close at hand and far beyond.

Voices from the Ice Edge

Little Diomede Island, Bering Strait, July 1998
We squint through the fog, knowing full well it is there. The charts say so, and its outline is showing clearly now on the radar. But to our eyes there is nothing—no rock, no coast, no sign that anything, or anyone, is in our ship's path. Slowly, a vague form starts to force its way through the mist, but as the image becomes gradually clearer, our blindness gives way to incredulity. It seems almost impossibly bleak, reaching out of the gray sea and disappearing into the clouds.

Within hours, though, the fog has cleared—for the first time, it seems, in days. This afternoon, at least, the Bering Strait is calm, and Little Diomede Island is in full view. Anthony Soolook Jr., Iñupiat fisherman, whaler, and walrus hunter, stands on the ship's deck and looks back at his home. The tiny village sits bravely in the path of the Arctic storms that regularly hurtle past.

The island is barren and littered with boulders, some of them looming ominously over the tiny village. Anthony points to places where the rocks have become dislodged from their perches, carving tracks through the shallow soil. "Our island is falling apart," he says matter-of-factly. "Last night I had a dream, that the boulders came down and smashed our village."

There have always been landslides on Little Diomede. The soil is thin and loose, providing little support for the rocks on the surface. But it generally has all been anchored in place by the permafrost. Now, however, temperatures are increasing, the permafrost is melting, the ground above it is sinking, and the landslides are becoming more frequent and more severe. Recently, a slide destroyed most of a carefully constructed pathway up the island. "In the past, when that happened, we would rebuild the path," said one village elder. "Now, I don't think we will bother."

Melting permafrost and falling rock are not the only threat to Little Diomede, nor is the village the only one whose inhabitants have noticed changes in their environment. Along the Bering and Chukchi Sea coasts of Alaska, from Gambell and Savoonga on remote St. Lawrence Island to the northern whaling towns of Wainwright and Point Lay, villagers tell strikingly similar stories: of ice forming later in the fall, breaking up sooner in the spring, and generally being thinner and less stable than usual. Of changes in the migration patterns of some species and of the appearance of other species totally new to the area. Of landslides and erosion, villages in danger of being buried or swept away. Of a long-standing, traditional subsistence way of life that is in danger of dying out.

In 1997 and 1998, the Greenpeace ship *Arctic Sunrise* visited Yup'ik and Iñupiat settlements; at a series of town meetings and in one-on-one interviews, researchers and ships' crew members gathered testimonies from villagers about the changes they are seeing in their climate and the world around them.

Jimmie Toolie is Savoonga's eldest elder, a distinction that grants him considerable authority in a culture in which age and experience are afforded a great deal of respect. Unlike some younger Alaska Natives, who watch television, speak English, and worship Michael Jordan, Jimmie is comfortable only with his native Yup'ik tongue. As he talks, slowly, his words are translated by another villager.

"There used to be heavy snowfall in the spring time; there used to be three feet of slush where we walked and now I don't see it any more," he says. "Instead of dog mushing we had dog slushing."

A younger villager, a hunter by the name of Benjamin Pungowiyi, chimes in:

> We didn't have much snow this year. When we go to our camp we have to cross these mountains. People had to pack up right away and come home because the trail conditions were really deteriorating—lack of snow, more bare rocks. As far as freezing up, we hardly got any snow until November. Usually we have our first snowfall around the end of September. During the summer months we have clouds and rain and drizzle. Now there's hardly any clouds or rain or drizzle, there's more sunshine. It's a lot warmer than before.

As with snow, so with ice. What scientists have predicted and observed, Alaska's coastal peoples are seeing come to pass. Perhaps more than any other observation, residents of the villages visited by Greenpeace expressed concern over changes in sea-ice cover. "The most change I've seen is how thin the ice is getting. Year by year," notes Benjamin Pungowiyi. Explains Stanley Oxereok, a middle-aged man with a moustache and a ready smile, sporting a Houston Rockets cap and standing on the beach at Wales: "The ice used to be five to six feet thick. The last couple of years it's been four, four and a half feet. That's a foot, a foot and a half, and that's a pretty substantial difference. . . . Break up seems to come quicker. Sometimes a couple of weeks, sometimes as much as a month sooner. . . . Freeze up was as much as a month late."

For Yup'ik and Iñupiat communities, changes in the extent and thickness of sea ice are of more than passing concern.

"It makes it hard to hunt in fall time when the ice starts forming," says Benjamin Neakok, an elder from Point Lay with proud features and thinning white hair. "It's kind of dangerous to be out. It's not really sturdy. And after it freezes there's always some open spots." In 1998, whalers from the town of Wainwright, on Alaska's North Slope, had to be rescued after the ice broke apart around their camp, leaving them drifting out to sea on an ice floe.

Gathering food on land is also, say many villagers, becoming increasingly difficult as conditions change. "We've really been hurting for berries the last three years but this is the worst," says Gail Moto, a native of the mainland United States who married an Iñupiaq and moved to Deering. "We knew that was going to happen because the elders know that the rain is connected to the berries, and they know if there is no rain, the berries are going to be poor. There's been less and less rain." Adds Kotzebue resident Hannah Mendenhall: "The thing that I notice when I walk out on the tundra, I notice that the tundra itself is not as spongy as it used to be. Now I can hear it crackle when I walk on it, and it's dry. It's real dry. Whereas before some places I don't go to because they're too wet, now [in] the areas that used to be lakes, all the plants are dried."

For subsistence cultures, the long-term consequences of such changes could be severe. Their way of life is already under assault, with cultural influences from television, video, and Western religion and education intruding on traditional values. Now, fear some, if there are fewer opportunities to go out and hunt or gather berries, those cultures that propagate themselves through observation and participation may be in danger of dying out entirely.

Furthermore, a changing climate may put the very existence of some villages at risk. In addition to the landslides threatening Little Diomede, surging storms—which are predicted to increase as a result of climate change—cause erosion, imperiling some villages built on the coast. This problem is further exacerbated by reduced shore ice, which normally protects the land from the storms' worst effects. Gambell, located on a narrow gravel spit, is repeatedly being moved farther inland as the spit steadily erodes. In 1997, a series of huge storms devastated the village of Shishmaref, taking out the seawall and nearly destroying several houses. After spending much of 1998 attempting to rebuild their island home, the villagers were forced to move several houses to a new location.

❋ ❋ ❋

On Little Diomede Island, the day is coming to a close. For the Greenpeace crew, it has been a long, hard one. The children of the village, unaccustomed to visitors of any stripe—let alone ones disgorged from a big green ship, dressed in bright red survival suits, and zipping back and forth between ship and shore in speedy inflatable boats—have followed them through the village endlessly and noisily. For their benefit, the activists have pretended to be steam engines, chugging and choo-chooing their way across the island as a trainload of young Iñupiat Eskimos snaked behind them. That done, they have dutifully swallowed mugs of stinkweed tea offered up to them by the children's grateful parents. And as the day has drawn on, the wind and waves have increased once again, so boat trips from the *Arctic Sunrise* to the island have drenched them from head to foot.

Now the village meeting itself is coming to an end, and as it does so, the accommodating villagers treat their visitors to an evening of Eskimo dancing—traditional theater in which men and women alike enact scenes of hunting and fishing. Among the performers is a young man, maybe nineteen, with the first hints of a moustache penciled along his upper lip. He is dressed like any American of his age: baggy pants, hooded sweatshirt, a gold necklace. But soon he is deep into his dancing, acting out the hunting of a walrus with a power and energy that has his spectators on the edge of their seats. He is a classic example of the dilemma facing modern-day Eskimos: of "walking between two worlds," embracing modern culture while retaining traditional values. That night on Little Diomede, with the rocks looming overhead and the wind howling outside, it was hard not to wonder whether that traditional style of life—or even the village itself—could survive much longer.

Epilogue

The story comes, finally, full circle. The first sustained European interest in the polar regions centered on a belief in their potential as a passage to the East and a belief in the existence of trade routes via the northeast and northwest. Repeated encounters with the Arctic's true nature mostly disabused people of those notions; for the best part of three centuries, searches for a Northwest Passage and a Northeast Passage were challenges of exploration rather than commerce, and only an occasional effort, such as that of the would-be oil tanker *Manhattan* in 1969, sought seriously to test the financial viability of a pathway through the ice.

Now, however, the issue is once more on the table. If rising temperatures are diminishing the thickness and extent of Arctic sea ice, as appears to be the case, routinely navigable Northeast and Northwest Passages may finally become a reality, allowing the fulfillment of a goal several centuries in the making. There are those who are actively counting on it. In the fall of 1999, government ministers and shipping executives gathered in Oslo, Norway, to discuss the potential profits to be gained from a shipping lane between western Europe and Asia via the seas of the Russian Arctic. At about the same time, a Russian tug, the *Irbis,* hauled a huge floating dry dock from the Kamchatka Peninsula east through the Northwest Passage to Freeport, Bahamas. That summer, a Chinese government vessel appeared as if from nowhere to berth at the tiny, remote fishing port of Tuktoyaktuk in the northern Northwest Territories. A few crew members disembarked and took photographs until the local constable began asking questions, at which point they reboarded and the ship steamed away.

Thus far, at least, Russia seems enthusiastic about the potential benefits of a navigable Northeast Passage; when it comes to the Northwest Passage, however, Canada appears deeply conflicted. Simultaneously nervous that regular traffic would inevitably lead to oil spills and anxious to ensure that Canada will be the country to benefit politically and financially in the event that a truly navigable route ever does emerge through its Arctic archipelago, Ottawa claims the Northwest Passage as an internal waterway. Other countries, most conspicuously and vociferously the United States, dispute the Canadian assertion and insist that the passage is—or, perhaps more precisely, will be—an international right-of-way.

✳ ✳ ✳

Throughout much of the history of human involvement in the Arctic and Antarctic, both regions have proven attractive because of the "resources" they contained, such as whales and fur seals, which had already been massively depleted elsewhere, or which—in the case of oil—provided the opportunity to continue the expansion of industries already well under way in other parts of the world. They were, in each case, among the final frontiers the Western world breached.

Even as their secrets were slowly peeled away, the poles remained isolated and largely untouched. Alternating waves of explorers, whalers, sealers, and others appeared from over the horizon, did what they had to do, and then went home. No longer: today the Antarctic boasts human visitors year-round; after being sparsely inhabited for centuries, the Arctic is now extensively so; and both regions feel the effects of human activities far beyond their boundaries. Arctic natives have higher levels of contaminants in their bodies than do people anywhere else in the world. Ozone levels are at their lowest over the poles. Some of the earliest and most dramatic evidence of climate change is being seen in the Arctic and Antarctic. The polar regions, having been among the final places to suffer from human attention, are now among the first to show the warning signs of the consequences of activities conducted thousands of miles away. It remains to be seen whether we will be wise enough to pay attention.

Even so, the polar realms remain—at least in comparison with much of the rest of the globe—relatively unspoiled. But the fact that they have persisted in that state is no guarantee they will continue that way. It is possible that in the middle of the twenty-first century there will be oil drilling in Antarctica, the Northwest Passage will be one of the world's most important commercial shipping routes, and neither region will look much as it does today. Alternatively, a developing environmental ethic may prolong protection of the Antarctic and bring about a reduction in the drilling for fossil fuels in the Arctic, reducing the risk of pollution and putting the brakes on global warming. Or perhaps some other, unrelated development, completely unanticipated in the preceding pages, will emerge, thrusting the polar regions either to the forefront or further to the rear of human concerns.

One thing, however, seems certain. For eons the polar regions existed, through stasis and change, without any hint of human involvement or even existence. Similarly, the bulk of human history has unfolded with the great majority of its participants unfamiliar with or completely unaware of the Arctic and Antarctic. But as the world grows ever smaller, the ends of the Earth are no longer out of reach. In the course of the twenty-first century, the poles will continue to be ever more accessible, and potentially more vulnerable, than ever before.

Notes

Prologue

p. 1 **"beyond the north wind," "feasts out of sheer joy":** The Northern Lights Route, University Library of Tromsø, available on-line at http://www.ub. uit.no/northernlights/eng/pytheas.htm.

p. 2 **"the seas about the region":** Holland, *Arctic Exploration and Development, c. 500 B.C. to 1915,* p. 1.

p. 3 **"struck cold weather":** Owen, *The Antarctic Ocean,* p. 7.

p. 3 **"the vast pit of the abyss":** Vaughan, *The Arctic,* p. 41.

p. 4 **"prodigious seas," "misty weather," "islands of ice," "had not the sight of one fish," "not doubting":** Owen, *The Antarctic Ocean,* pp. 8–9.

p. 5 **"that reservoir of frost and snow":** Spufford, *I May Be Some Time,* p. 12.

p. 5 **"I am just going outside":** Scott, *Scott's Last Expedition,* p. 430.

p. 6 **"this journey":** Scott, *Scott's Last Expedition,* p. 442.

Chapter I: Poles Apart

p. 13 **"can be strong enough to warm rocks":** Stonehouse, *North Pole, South Pole,* p. 27.

p. 15 **"than can be found in a small garden":** Campbell, *The Crystal Desert,* p. 27.

p. 15 **"volcanic Zavadovskiy Island":** Campbell, *The Crystal Desert,* p. 27.

p. 21 **"bare and desolate waste":** Bruemmer et al., *The Arctic World,* p. 19.

p. 21 **"lifeless, except for millions of caribou":** Bruemmer et al., *The Arctic World,* p. 19.

p. 22 **"Southerners commonly think":** Bruemmer et al., *The Arctic World,* p. 13.

p. 22 **"a part of Scandinavia so warmed":** Lopez, *Arctic Dreams,* p. 18.

Chapter 2: Hunting the Bowhead

p. 35 **"the greatest sixteenth-century Arctic explorer," "outstanding navigational skills":** Vaughan, *The Arctic,* p. 59.

p. 35 **"incapable of fatigue":** Vaughan, *The Arctic,* p. 59.

p. 35 **"made all the hairs of our heads to rise":** Mirsky, *To the Arctic!* p. 39.

p. 35 **"in great cold, poverty, misery, and grief":** Mirsky, *To the Arctic!* p. 39.

p. 35 **"for there was none growing upon that land":** Mirsky, *To the Arctic!* p. 39.

p. 36 **"were taken with a sudden dizziness," "the cold, which before had been so great an enemy":** Mirsky, *To the Arctic!* p. 41.

p. 37 **"the sun at the far north":** Mirsky, *To the Arctic!* p. 44.

p. 38 **"in any seas whatsoever":** Vaughan, *The Arctic,* p. 77.

p. 39 **"that she upset and was lost":** Ellis, *Men and Whales,* p. 57.

p. 42 **"The whales, which were so constantly and vigorously pursued":** Scoresby, *An Account of the Arctic Regions, with a Description and History of the Northern Whale-Fishery,* vol. 2, p. 56.

p. 42 **"discovery for discovery's sake ceased":** Mirsky, *To the Arctic!* p. 45.

p. 45 **"I then saw lying upon the deck a finger":** Francis, *A History of World Whaling,* p. 169.

p. 46 **"There is heavy responsibility . . . officers and crew will unite to put him down":** Bockstoce, *Whales, Ice, and Men,* p. 23.

p. 46 **"I actually believe if they had any hope":** Bockstoce, *Whales, Ice, and Men,* p. 24.

p. 46 **"new-fangled monster":** Bockstoce, *Whales, Ice, and Men,* p. 24.

p. 46 **"went down and ran along the bottom":** Bockstoce, *Whales, Ice, and Men,* p. 24.

p. 50 **"thick as bees":** Bockstoce, *Whales, Ice, and Men,* p. 256.

p. 52 **"I really can't recall any disagreements":** Eber, *When the Whalers Were Up North,* p. 165.

p. 52 **"Plenty of native company":** Bockstoce, *Whales, Ice, and Men,* p. 181.

p. 53 **"buy a hundred dollars' worth of bone":** Bockstoce, *Whales, Ice, and Men,* p. 184.

p. 53 **"drank excessively from the first drop":** Ray, *The Eskimos of Bering Strait, 1650–1898,* p. 252.

p. 53 **"will take nothing else if they can get rum"**: Bockstoce, *Whales, Ice, and Men,* p. 184.

p. 53 **"demand [was] usually for rum . . . so the influence could be better"**: Bockstoce, *Whales, Ice, and Men,* p. 182.

p. 54 **"barrels and barrels"**: Boeri, *People of the Ice Whale,* p. 56.

p. 54 **"islanders drank to the point of oblivion"**: Boeri, *People of the Ice Whale,* p. 56.

p. 54 **"ships' riflemen may have gunned down 200,000 walrus"**: Boeri, *People of the Ice Whale,* p. 55.

p. 57 **"in the tens"**: Burns, Montague, and Cowles, *The Bowhead Whale,* p. 478.

Chapter 3: Terra Incognita

p. 63 **"an obdurate, cantankerous Scot"**: Mickleburgh, *Beyond the Frozen Sea,* p. 16.

p. 63 **"fiery-tempered," "too often ungracious and ill-natured," "he expressed his views without reserve"**: Mill, *The Siege of the South Pole,* p. 57.

p. 63 **"a Scot with a grievance"**: Gurney, *Below the Convergence,* p. 16.

p. 63 **"No contemporary would deny," "would deny his ability"**: Beaglehole, *The Life of Captain James Cook,* p. 104.

p. 64 **"a greater extent than the whole civilised part of Asia," "it rests to shew"**: Beaglehole, *The Life of Captain James Cook,* p. 121.

p. 65 **"into all the world, and preach the gospel"**: The Holy Bible, Matthew 28:19.

p. 65 **"it would be admitting the existence of souls"**: Mill, *The Siege of the South Pole,* p. 14.

p. 65 **"Can anyone be so foolish"**: Daniel J. Boorstin, *The Discoverers: A History of Man's Search to Know His World and Himself* (New York: Random House, 1983), p. 107.

p. 66 **"The Antarctic problem"**: Mill, *The Siege of the South Pole,* pp. 13–14.

p. 67 **"asked nothing but to lead a life of contentment"**: Reader's Digest, *Antarctica,* p. 70.

p. 67 **"not in the possession of any Christian Prince"**: Martin, *A History of Antarctica,* p. 35.

p. 68 **"the winds were such"**: Francis Fletcher, *The World Encompassed by Sir Francis*

Drake, Collected Out of the Notes of Master Francis Fletcher (London: Nicholas Bourne, 1652), p. 42.

p. 68 **"southernmost man in the world"**: Gurney, *Below the Convergence,* p. 9.

p. 68 **"not to sale any further toward the pole Antarctick"**: Martin, *A History of Antarctica,* p. 36.

p. 69 **"hospitable, civil and ingenious peoples"**: Mickleburgh, *Beyond the Frozen Sea,* p. 16.

p. 69 **"There is at present no trade"**: Beaglehole, *The Life of Captain James Cook,* p. 121.

p. 71 **"a proper person to send to the South Seas"**: Beaglehole, *The Life of Captain James Cook,* p. 102.

p. 71 **"Wherever I am in June 1769"**: Beaglehole, *The Life of Captain James Cook,* p. 103.

p. 72 **"may be necessary to observe . . . intended for the Service"**: Beaglehole, *The Life of Captain James Cook,* p. 103.

p. 72 **"proceed round Cape Horn," "When this Service is perform'd"**: Beaglehole, *The Life of Captain James Cook,* p. 147.

p. 72 **"When this Service is perform'd"**: Beaglehole, *The Life of Captain James Cook,* p. 148.

p. 73 **"Whereas the making Discoverys"**: Beaglehole, *The Life of Captain James Cook,* pp. 148–149.

p. 73 **"with the Consent of the Natives"**: Beaglehole, *The Life of Captain James Cook,* p. 148.

p. 74 **"the health of her Crew"**: Beaglehole, *The Life of Captain James Cook,* p. 149.

p. 74 **"I am very far from intending . . . I would *not have come back in Ignorance*"**: Mill, *The Siege of the South Pole,* p. 59.

p. 75 **"That a Southern Continent exists"**: Joseph Banks, *The Endeavour Journal of Joseph Banks, 1768–1771,* vol. 2 (Sydney: Angus & Robertson, 1962), p. 40.

p. 75 **"I think it would be a great pitty"**: Beaglehole, *The Life of Captain James Cook,* p. 278.

p. 76 **"Upon due consideration of the discoveries"**: Beaglehole, *The Life of Captain James Cook,* pp. 286–287.

p. 77 **"If land is discovered"**: Beaglehole, *The Life of Captain James Cook,* p. 287.

p. 78 **"The Clowds near the horizon"**: Cook, *The Journals of Captain James Cook on His Voyages of Discovery,* vol. 2, p. 321.

p. 78 **"I will not say it was impossible"**: Cook, *A Voyage Towards the South Pole and Around the World,* vol. 1, p. 268.

p. 79 **"The head of the bay"**: Cook, *A Voyage Towards the South Pole and Around the World,* vol. 2, p. 213.

p. 79 **"The disappointment I now felt"**: Cook, *The Journals of Captain James Cook on His Voyages of Discovery,* vol. 2, p. 636.

p. 79 **"I firmly believe"**: Cook, *A Voyage Towards the South Pole and Around the World,* vol. 2, p. 231.

p. 80 **"The shores . . . swarmed with young cubs"**: Cook, *A Voyage Towards the South Pole and Around the World,* vol. 2, p. 213.

p. 81 **"in the manner of salted dried cod fish"**: Gurney, *Below the Convergence,* p. 149.

p. 81 **"almost entirely abandoned by the animals"**: Gurney, *Below the Convergence,* p. 149.

p. 81 **"these animals [were] now nearly extinct"**: Weddell, *A Voyage Towards the South Pole,* p. 53.

p. 83 **"vast quantities of seals, whales"**: Gurney, *Below the Convergence,* p. 156.

p. 83 **"a prospect the most gloomy . . . the long-sought Southern Continent"**: Gurney, *Below the Convergence,* p. 160.

p. 85 **"we encountered icebergs"**: Bellingshausen, *The Voyage of Captain Bellingshausen to the Antarctic Seas, 1819–1821,* vol. 1, p. 117.

p. 86 **"The harvest of the seas"**: Mickleburgh, *Beyond the Frozen Sea,* p. 31.

Chapter 4: So Remorseless a Havoc

p. 89 **"may it continue to the Pole"**: Ross, *Ross in the Antarctic,* p. 82.

p. 90 **"way of restoring to England the honour"**: Ross, *A Voyage of Discovery and Research in the Southern and Antarctic Regions During the Years 1839–1843,* vol. 1, p. 188.

p. 90 **"extreme solitude and omnipotent grandeur"**: Ross, *Ross in the Antarctic,* p. 88.

p. 90 **"The volumes of smoke"**: Ross, *Ross in the Antarctic,* pp. 91–92.

p. 91 **"This was a sight so surpassing everything"**: Ross, *Ross in the Antarctic,* p. 92.

p. 91 **"As we approached the land"**: Ross, *A Voyage of Discovery and Research in the Southern and Antarctic Regions During the Years 1839–1843,* vol. 1, pp. 217–219.

p. 94 **"A great many whales were observed"**: Ross, *A Voyage of Discovery and Research,* vol. 1, pp. 191–192.

p. 95 **"We observed a very great number of the largest-sized black whales"**: Ross, *A Voyage of Discovery and Research in the Southern and Antarctic Regions During theYears 1839–1843,* vol. 2, p. 327.

p. 96 **"The *Balaena mysticetus*"**: Murdoch, *From Edinburgh to the Antarctic,* pp. 23–24.

p. 97 **"saw any sign of a whale"**: Murdoch, *From Edinburgh to the Antarctic,* p. 357.

p. 97 **"The whale went off in a bee-line"**: Murdoch, *From Edinburgh to the Antarctic,* p. 291.

p. 100 **"I was sitting foremost in the boat"**: Rosove, *Let Heroes Speak,* p. 65.

p. 100 **"I do not know whether it was the desire"**: Rosove, *Let Heroes Speak,* p. 65.

p. 100 **"proved that landing on Antarctica proper"**: Reader's Digest, *Antarctica,* p. 129.

p. 101 **"consist of at least two vessels"**: Reader's Digest, *Antarctica,* p. 129.

p. 101 **"further exploration of the Antarctic regions"**: Reader's Digest, *Antarctica*, p. 130.

p. 102 **"Two days later"**: Nordenskjöld, *Antarctica,* p. 393.

p. 103 **"The days came and the days went"**: Nordenskjöld, *Antarctica,* p. 436.

p. 104 **"We soon reach the cape mentioned"**: Nordenskjöld, *Antarctica,* pp. 307–308.

p. 105 **"For the second time within a month"**: Nordenskjöld, *Antarctica,* p. 504.

p. 106 **"None of us intended going to bed that night"**: Nordenskjöld, *Antarctica,* pp. 508–510.

p. 107 **"highly original English"**: Tønnessen and Johnsen, *The History of Modern Whaling,* p. 160.

p. 107 **"I tank youse"**: Tønnessen and Johnsen, *The History of Modern Whaling,* p. 160.

p. 108 **"surpassed anything ever seen"**: Tønnessen and Johnsen, *The History of Modern Whaling,* p. 183.

p. 109 **"a qualified failure"**: Ellis, *Men and Whales,* p. 359.

p. 111 **"the interest of the nations of the world," "the history of whaling," "to provide for the proper conservation," "thus make possible the orderly development"**: Preamble, International Convention for the Regulation of Whaling, Washington, D.C., 1946.

p. 112 **"Conscious of the importance"**: Friends of the Earth, *The Whale Manual* (San Francisco: Friends of the Earth, 1978), p. 25.

p. 113 **"it may soon become necessary"**: Friends of the Earth, *The Whale Manual,* p. 21.

p. 113 **"powers of recovery"**: Friends of the Earth, *The Whale Manual,* p. 21.

p. 116 **"the most sordid, unsanitary habitation"**: Ellis, *Men and Whales,* p. 352.

Chapter 5: The Last Wilderness

p. 117 **"as if the steamers appeared in the bay one day"**: Fox, *Antarctica and the South Atlantic,* p. 36.

p. 119 **"would risk provoking"**: Freedman and Gamba-Stonehouse, *Signals of War,* p. 43.

p. 120 **"fresh water in plenty," "abundance of geese and ducks," "as for wood"**: Hastings and Jenkins, *The Battle for the Falklands,* p. 2.

p. 120 **"even in time of peace . . . make us master of the seas"**: Hastings and Jenkins, *The Battle for the Falklands,* p. 2.

p. 121 **"I tarry on this miserable desert"**: Hastings and Jenkins, *The Battle for the Falklands,* p. 3.

p. 121 **"thrown aside from human use"**: Hastings and Jenkins, *The Battle for the Falklands,* p. 4.

p. 122 **"to restore the King's honor"**: Hastings and Jenkins, *The Battle for the Falklands,* p. 4.

p. 122 **"and the islands adjacent"**: Hastings and Jenkins, *The Battle for the Falklands,* p. 4.

p. 123 **"that little ice-cold bunch of land"**: Freedman and Gamba-Stonehouse, *Signals of War,* p. 154.

p. 123 **"freak of history," "even after men began to die"**: Hastings and Jenkins, *The Battle for the Falklands,* p. vii.

p. 124 **"HMG [His Majesty's Government] have under consideration"**: Dodds, *Geopolitics in Antarctica,* pp. 79–80.

p. 124 **"the time has not yet arrived"**: Dodds, *Geopolitics in Antarctica,* p. 80.

p. 126 **"that part of His Majesty's Dominions," "to make all such Rules"**: Quartermain, *New Zealand and the Antarctic,* p. 40.

p. 126 **"the preference of the British government"**: Dodds, *Geopolitics in Antarctica,* p. 82.

p. 127 **"formed by all lands":** Dodds, *Geopolitics in Antarctica,* p. 112.

p. 127 **"Chile must call attention to herself":** Dodds, *Geopolitics in Antarctica,* p. 113.

p. 127 **"southern seas of the republic":** Dodds, *Geopolitics in Antarctica,* p. 49.

p. 129 **"For a time several of the landing party were actively engaged":** Fuchs, *Of Ice and Men,* p. 27.

p. 130 **"as old as the nation itself":** Bertrand, *Americans in Antarctica, 1775–1948,* p. 18.

p. 131 **"a milestone in American science":** Bertrand, *Americans in Antarctica, 1775–1948,* p. 159.

p. 131 **"such a wealth":** Bertrand, *Americans in Antarctica, 1775–1948,* p. 159.

p. 134 **"that human beings can permanently occupy":** Quigg, *A Pole Apart,* p. 132.

p. 135 **"the discovery of lands unknown":** Shapley, *The Seventh Continent,* p. 43.

p. 135 **"consolidate and extend the basis for United States claims":** Bertrand, *Americans in Antarctica, 1775–1948,* p. 484.

p. 135 **"for my country, so far as this act allows":** Bertrand, *Americans in Antarctica, 1775–1948,* p. 403.

p. 135 **"establishing and strengthening US claims," "take any action . . . United States government," "United States sovereignty":** Joyner and Theis, *Eagle over the Ice,* pp. 38–39.

p. 136 **"some study to a new form of sovereignty":** Quigg, *A Pole Apart,* p. 132.

p. 137 **"indisputable right of the Soviet Union":** Quigg, *A Pole Apart,* p. 135.

p. 141 **"we do not insist":** Shapley, *The Seventh Continent,* p. 85.

p. 141 **"What do you think of it . . . what use is it?":** Siple, *Ninety Degrees South,* p. 89.

p. 141 **"Good. Let's get the hell out of here":** Dufek, *Operation Deepfreeze,* p. 201.

p. 143 **"It is in our interest to insure":** Quigg, *A Pole Apart,* p. 144.

p. 143 **"At the frozen bottom of the earth":** Dodds, *Geopolitics in Antarctica,* p. 36.

p. 143 **"free trade in science," "should not be . . . used for military purposes":** Quigg, *A Pole Apart,* p. 143.

p. 144 **"the interests of mankind":** Quigg, *A Pole Apart,* p. 143.

p. 144 **"Antarctica shall be used,"** Antarctic Treaty, Washington, D.C., 1959, Article I.1.

p. 144 **"nuclear explosions in Antarctica,"** Antarctic Treaty, Article V.1.

p. 144 **"designate observers,"** Antarctic Treaty, Article V.1.

p. 144 **"complete freedom of access,"** Antarctic Treaty, Article V.2.

p. 145 **"No acts or activities,"** Antarctic Treaty, Article IV.2.

p. 145 **"in the interest of all mankind":** Antarctic Treaty, Preamble.

p. 146 **"last till a big mineral discovery is made":** Suter, *Antarctica,* p. 46.

p. 150 **"urge their nationals and other States to refrain":** Shapley, *The Seventh Continent,* p. 139.

p. 150 **"for the benefit of mankind as a whole":** Joyner, *Governing the Frozen Commons,* p. 237.

p. 151 **"no Antarctic mineral resource activity will be permitted":** Convention on the Regulation of Antarctic Mineral Resource Activities (CRAMRA), Wellington, New Zealand, 1988, Article 3.

p. 151 **"damage to the Antarctic environment," "beyond that which is negligible":** CRAMRA, Article 1.15.

p. 153 **"the majority of existing research stations," "radical modification":** May, *The Greenpeace Book of Antarctica,* p. 132.

p. 154 **"one of the most heavily polluted":** Greenpeace, *State of the Ice: An Overview of the Human Impacts in Antarctica* (Amsterdam: Greenpeace, 1994), p. 7.

p. 157 **"natural reserve, devoted to peace and science,"** Protocol on Environmental Protection to the Antarctic Treaty, Madrid, Spain, 1991, Article 2.

p. 158 **"The protection of the Antarctic environment,"** Environmental Protocol, Article 3.1.

p. 158 **"Any activity relating to mineral resources":** Environmental Protocol, Article 3.1.

Chapter 6. Crude Awakening

p. 161 **"encumbered, some of them jammed and crowded":** A. W. Greely, *Handbook of Alaska* (New York: Scribner, 1909), p. 156.

p. 162 **"along the shore that runs to the north," "where it joins with America," "sail to some settlement . . . bring it here":** Mirsky, *To the Arctic!* p. 71.

p. 163 **"open up all of northeastern Asia":** Ford, *Where the Sea Breaks Its Back,* pp. 49–50.

p. 166 **"We now found ourselves . . . without hindrance," "most delicious,"** "beyond comparison . . . almonds," "exceedingly savory": Ford, *Where the Sea Breaks Its Back*, pp. 160–161.

p. 170 "the most illiterate": Cook, *The Journals of Captain James Cook on His Voyages of Discovery*, vol. 3, p. 456.

p. 171 "the rage with which our seamen were possessed": McCracken, *Hunters of the Stormy Sea*, p. 104.

p. 173 "oil will be found in Alaska": Naske and Slotnick, *Alaska*, p. 243.

p. 176 "the Prudhoe well had been dry": Roderick, *Crude Dreams*, p. 213.

p. 176 "It looks extremely good . . . We have a major discovery": Strohmeyer, *Extreme Conditions*, p. 58.

p. 177 "'It's another Kuwait' . . . they just may be right": Chasan, *Klondike '70*, p. 4.

p. 177 "timidity," "a little bit afraid of bigness," "get Alaska moving": Coates, *The Trans-Alaska Pipeline Controversy*, p. 162.

p. 177 "opening up": Coates, *The Trans-Alaska Pipeline Controversy*, p. 163.

p. 177 "forever banish the myth": Coates, *The Trans-Alaska Pipeline Controversy*, p. 163.

p. 177 "the harsh kind of region": Coates, *The Trans-Alaska Pipeline Controversy*, p. 164.

p. 178 "This impossible road . . . one of its greatest assets": Roderick, *Crude Dreams*, p. 256.

p. 178 "So they've scarred the tundra": Coates, *The Trans-Alaska Pipeline Controversy*, p. 167.

p. 179 "ugly symptom," "the first violent change," "end of thousands of years of solitude": Coates, *The Trans-Alaska Pipeline Controversy*, p. 166.

p. 179 "Vessel in sight, Sir": Savours, *The Search for the Northwest Passage*, p. 308.

p. 181 "He faced narrow waters": Huntford, *The Last Place on Earth*, p. 83.

p. 181 "big enough to drive a truck through": Naske and Slotnick, *Alaska*, p. 256.

p. 181 "a fleet of trucks in permanent circulation": Yergin, *The Prize*, p. 572.

p. 182 "The line will reach altitudes of 4,700 feet": Rogers, *Change in Alaska*, p. 10.

p. 183 "I saw thousands of oil cans": Rogers, *Change in Alaska*, pp. 173–174.

p. 183 "In North America, the petroleum industry . . . in this new and severe environment": Rogers, *Change in Alaska*, p. 151.

p. 183 "run through a bleak, barren, hostile, unpopulated country," "the most desolate area in the world": Coates, *The Trans-Alaska Pipeline Controversy*, p. 204.

p. 184 "float," "we could expect to see buried pipe": Coates, *The Trans-Alaska Pipeline Controversy*, p. 183.

p. 184 "an ecological Berlin Wall": Rogers, *Change in Alaska*, pp. 11–12.

p. 184 "In the Arctic region, caribou normally spend the winters": Chasan, *Klondike '70*, p. 133.

p. 184 "God placed these things beneath the surface": Strohmeyer, *Extreme Conditions*, p. 84.

p. 184 "I am up to here with people": Coates, *The Trans-Alaska Pipeline Controversy*, p. 180.

p. 184 "If you took a daily newspaper": Coates, *The Trans-Alaska Pipeline Controversy*, p. 202.

p. 184 "If oil is pumped out of Prudhoe Bay": Strohmeyer, *Extreme Conditions*, p. 83.

p. 184 "The fears about damage": Strohmeyer, *Extreme Conditions*, p. 84.

p. 185 "there will be not one drop of oil": Strohmeyer, *Extreme Conditions*, p. 84.

p. 185 "until the discovery of oil in 1957," "We have always known": Roderick, *Crude Dreams*, p. 293.

p. 185 "there are no living organisms": Coates, *The Trans-Alaska Pipeline Controversy*, p. 180.

p. 186 "How can white men sell our land . . . let the pigs pay the rent": Roderick, *Crude Dreams*, p. 294.

p. 186 "neither the United States": Naske and Slotnick, *Alaska*, pp. 201–202.

p. 186 "just because somebody's grandfather chased a moose": Strohmeyer, *Extreme Conditions*, p. 70.

p. 187 "the fundamental change that development would bring," "the original character," "adaptable": Coates, *The Trans-Alaska Pipeline Controversy*, p. 196.

p. 187 "it was not written by a proponent": Coates, *The Trans-Alaska Pipeline Controversy*, p. 228.

p. 188 "I cannot get overly upset": Coates, *The Trans-Alaska Pipeline Controversy*, p. 242.

p. 188 **"To preserve the 7,680 acres":** Coates, *The Trans-Alaska Pipeline Controversy*, p. 249.

p. 190 **"largest protected block of wild habitat in the world":** Miller, *Midnight Wilderness*, p. 225.

p. 193 **"We're playing Russian roulette":** Davidson, *In the Wake of the* Exxon Valdez, p. 9.

p. 194 **"I think we are in serious trouble":** Davidson, *In the Wake of the* Exxon Valdez, p. 18.

p. 194 **"We've fetched up," "Evidently we're leaking some oil":** Davidson, *In the Wake of the* Exxon Valdez, p. 19.

p. 196 **"no one can make a living":** Riki Ott, "To Those Living on Sound, Recovery Still Distant," *Anchorage Daily News*, March 8, 1999, B-8.

p. 196 **"If the water is dead. . . . but it's true":** Walter Meganack Sr., "Coping with the Time the Water Died," *Anchorage Daily News*, August 5, 1989, B-11.

Chapter 7: The Ends of the Earth

p. 201 **"We'll have to climb out of this":** Hickson, *Flight 901 to Erebus*, p. 11.

p. 202 **"members of the New Zealand parliament":** Hall and Johnston, *Polar Tourism*, p. 171.

p. 202 **"I didn't just read . . . I was *there* with them":** Lindblad and Fuller, *Passport to Anywhere*, p. 26.

p. 202 **"were concentrating on only a tired handful of countries":** Lindblad and Fuller, *Passport to Anywhere*, p. 28.

p. 203 **"a trickle of requests," "wanted to reach out to more exciting places":** Lindblad and Fuller, *Passport to Anywhere*, p. 28.

p. 203 **"botanical and horticultural wonders of the world":** Lindblad and Fuller, *Passport to Anywhere*, p. 37.

p. 203 **"the first tentative probe into Antarctica," "we were flooded with inquiries":** Lindblad and Fuller, *Passport to Anywhere*, p. 94.

p. 204 **"important that lots of us go":** Soper, *Antarctica*, p. 7.

p. 208 **"dead forest":** Arctic Monitoring and Assessment Programme, *Arctic Pollution Issues*, p. 117.

p. 209 **"a slightly flawed planet":** O'Neill, *The Firecracker Boys*, p. 25.

p. 213 **"the coniferous forest appears healthy":** Arctic Monitoring and Assessment Programme, *Arctic Pollution Issues*, p. 129.

p. 213   **"industrial deserts"**: Arctic Monitoring and Assessment Programme, *Arctic Pollution Issues*, pp. 101–102.

p. 213   **"the ecosystems of at least five water bodies"**: Arctic Monitoring and Assessment Programme, *Arctic Pollution Issues*, p. 103.

p. 213   **"are 10–380 times background levels"**: Arctic Monitoring and Assessment Programme, *Arctic Pollution Issues*, p. 104.

p. 223   **"disturbing new evidence [that the hole was caused by CFCs]"**: Benedick, *Ozone Diplomacy*, p. 20.

p. 224   **"primed for ozone destruction"**: Gribbin, *The Hole in the Sky*, p. 184.

p. 227   **"dense plume of fragments," "In 25 years . . . never seen anything like it"**: "One Small Ice Shelf Dies, One Giant Iceberg Born," press release, British Antarctic Survey, February 27, 1995.

p. 227   **"clearly an escalation"**: "Antarctic Ice Shelves Breaking Up Due to Decades of Higher Temperatures," press release, National Snow and Ice Data Center (NSIDC), University of Colorado at Boulder/British Antarctic Survey, April 7, 1999.

p. 229   **"oceans would flood all existing port facilities"**: R. R. Revelle, quoted in Michael Oppenheimer, "Global Warming and the Stability of the West Antarctic Ice Sheet," *Nature* 393 (May 28, 1998): 325.

p. 231   **"among the first documented biological effects"**: Interview with the author, September 3, 1998. See also Kieran Mulvaney, "Can't Take the Heat," *New Scientist* (September 26, 1998): 12.

p. 236   **"is not how much the forests will be damaged"**: Interview with the author, June 28, 1998. See also Kieran Mulvaney, "Eaten Alive," *New Scientist* (July 18, 1998): 12.

p. 236   **"balance of evidence suggests a discernible human influence"**: Intergovernmental Panel on Climate Change, *Climate Change 1995: The Science of Climate Change.* Cambridge, England: Cambridge University Press, 1996, p. 4.

p. 237   **"longer and more closely-scrutinized temperature record," "stronger evidence for a human influence," "very unlikely," "the observed warming over the last 100 years"**: Intergovernmental Panel on Climate Change, Draft Summary for Policymakers of IPCC Working Group I, October 2000.

p. 238   **"Our island is falling apart . . . boulders came down and smashed our village"**: Conversation with the author, July 16, 1998.

p. 238   **"In the past, when that happened . . . Now, I don't think we will bother"**: Conversation with the author, July 15, 1998.

p. 239 "There used to be heavy snowfall . . . we had dog slushing": Gibson and Schullinger, *Answers from the Ice Edge,* p. 13.

p. 239 "We didn't have much snow": Gibson and Schullinger, *Answers from the Ice Edge,* p. 14.

p. 239 "The most change I've seen": Gibson and Schullinger, *Answers from the Ice Edge,* p. 15.

p. 239 "The ice used to be five to six feet thick": Gibson and Schullinger, *Answers from the Ice Edge,* p. 15.

p. 239 "It makes it hard to hunt . . . always some open spots": Gibson and Schullinger, *Answers from the Ice Edge,* p. 19.

p. 240 "We've really been hurting": Gibson and Schullinger, *Answers from the Ice Edge,* p. 25.

p. 240 "The thing that I notice when I walk out on the tundra": Gibson and Schullinger, *Answers from the Ice Edge,* p. 26.

Selected Bibliography

There are a great many books available on the Arctic and Antarctic—their natural and human history, exploration, exploitation, and politics, among other topics. The following selection is by no means comprehensive. It does not list all the materials I consulted in the course of writing this book; it does, however, include those that I have found especially valuable and that I consider likely to be of particular interest to those wishing to learn more about the topics covered in each chapter. Some are available in a great many editions; in some cases more recent editions are listed even though the relevant publications first appeared decades or even centuries earlier.

Prologue

Holland, Clive. *Arctic Exploration and Development, c. 500 B.C. to 1915.* New York: Garland, 1994.

Huntford, Roland. *The Last Place on Earth.* New York: Atheneum, 1985.

Owen, Russell. *The Antarctic Ocean.* New York: Whittlesey House, 1941.

Scott, Robert Falcon. *Scott's Last Expedition: The Journals.* New York: Carroll & Graf, 1996.

Spufford, Francis. *I May Be Some Time: Ice and the English Imagination.* New York: St. Martin's Press, 1997.

Vaughan, Richard. *The Arctic: A History.* Dover, N.H.: Alan Sutton, 1994.

Chapter I: Poles Apart

Bonner, W. N., and D. W. H. Walton, eds. *Key Environments: Antarctica.* Oxford: Pergamon Press, 1985.

Bruemmer, Fred, et al. *The Arctic World.* New York: Portland House, 1989.

Campbell, David G. *The Crystal Desert: Summers in Antarctica.* Boston: Houghton Mifflin, 1992.

Fothergill, Alistair. *The Natural History of Antarctica: Life in the Freezer.* New York: Sterling, 1995.

Laws, Richard. *Antarctica: The Last Frontier.* London: Boxtree, 1989.

Lopez, Barry. *Arctic Dreams: Imagination and Desire in a Northern Landscape.* New York: Scribner, 1986.

Pielou, E. C. *A Naturalist's Guide to the Arctic.* Chicago: University of Chicago Press, 1994.

Rowell, Galen. *Poles Apart: Parallel Visions of the Arctic and Antarctic.* Berkeley: University of California Press, 1995.

Rubin, Jeff. *Antarctica: A Lonely Planet Travel Survival Guide.* Oakland, Calif.: Lonely Planet, 1996.

Soper, Tony. *Antarctica: A Guide to the Wildlife.* Old Saybrook, Conn.: Globe Pequot Press, 1996.

Stewart, John. *Antarctica: An Encyclopedia.* Jefferson, N.C.: McFarland, 1990.

Stonehouse, Bernard. *North Pole, South Pole: A Guide to the Ecology and Resources of the Arctic and Antarctic.* London: Prion Press, 1990.

Swaney, Deanna. *The Arctic.* Melbourne: Lonely Planet, 1999.

Young, Steven B. *To the Arctic: An Introduction to the Far Northern World.* New York: Wiley, 1994.

Chapter 2: Hunting the Bowhead

Bockstoce, John R. *Whales, Ice, and Men: The History of Whaling in the Western Arctic.* Seattle: University of Washington Press, 1995.

Boeri, David. *People of the Ice Whale: Eskimos, White Men, and the Whale.* New York: Dutton, 1983.

Brody, Hugh. *Living Arctic: Hunters of the Canadian North.* Seattle: University of Washington Press, 1987.

Burns, John J., J. Jerome Montague, and Cleveland J. Cowles, eds. *The Bowhead Whale.* Lawrence, Kans.: Society for Marine Mammalogy, 1993.

Damas, David, ed. *Handbook of North American Indians.* Vol. 5, *Arctic.* Washington, D.C.: Smithsonian Institution Press, 1984.

Donovan, G. P., ed. *Aboriginal/Subsistence Whaling (with Special Reference to the Alaska and Greenland Fisheries).* Cambridge, England: International Whaling Commission, 1982.

Eber, Dorothy Harley. *When the Whalers Were Up North: Inuit Memories of the Eastern Arctic.* Boston: David R. Godine, 1989.

Ellis, Richard. *Men and Whales.* New York: Knopf, 1991.

Francis, Daniel. *A History of World Whaling.* New York: Viking, 1990.

Hess, Bill. *Gift of the Whale: The Iñupiat Bowhead Hunt, A Sacred Tradition*. Seattle: Sasquatch Books, 1999.

Mirsky, Jeannette. *To the Arctic! The Story of Northern Exploration from Earliest Times*. Chicago: University of Chicago Press, 1970.

Napoleon, Harold. *Yuuyaraq: The Way of the Human Being*. Fairbanks: Alaska Native Knowledge Network, 1996.

Nelson, Richard. *Hunters of the Northern Ice*. Chicago: University of Chicago Press, 1972.

Ray, Dorothy Jean. *The Eskimos of Bering Strait, 1650–1898*. Seattle: University of Washington Press, 1975.

Ross, W. Gillies. *Arctic Whalers, Icy Seas: Narratives of the Davis Strait Whale Fisheries*. Toronto: Irwin, 1985.

Scoresby, William. *An Account of the Arctic Regions, with a Description and History of the Northern Whale-Fishery*. 2 vols. Newton Abbot, Devon, England: David & Charles, 1969.

Vaughan, Richard. *The Arctic: A History*. Dover, N.H.: Alan Sutton, 1994.

Chapter 3: Terra Incognita

Beaglehole, J. C. *The Life of Captain James Cook*. London: A & C Black, 1974.

Bellingshausen, Thaddeus. *The Voyage of Captain Bellingshausen to the Antarctic Seas, 1819–1821*. 2 vols. Edited by Frank Debenham. London: Hakluyt Society, 1945.

Bonner, W. Nigel. *Seals and Man: A Study of Interactions*. Seattle: Washington Sea Grant Program, 1982.

Cook, James. *The Journals of Captain James Cook on His Voyages of Discovery*. 4 vols. Edited by J. C. Beaglehole. Woodbridge, Suffolk, England: Boydell & Brewer, 1999.

Cook, James. *A Voyage Towards the South Pole and Around the World*. 2 vols. Adelaide: Libraries Board of South Australia, 1970.

Gurney, Alan. *Below the Convergence: Voyages Toward Antarctica, 1699–1839*. New York: Norton, 1997.

Jones, A. G. E. *Antarctica Observed: Who Discovered the Antarctic Continent?* Whitby, North Yorkshire, England: Caedmon of Whitby, 1982.

Martin, Stephen. *A History of Antarctica*. Sydney: State Library of New South Wales Press, 1996.

Mickleburgh, Edwin. *Beyond the Frozen Sea: Visions of Antarctica*. London: Bodley Head, 1987.

Mill, Hugh Robert. *The Siege of the South Pole*. London: Alston Rivers, 1905.

Reader's Digest. *Antarctica: The Extraordinary History of Man's Conquest of the Frozen Continent*. Surry Hills, New South Wales, Australia: Reader's Digest, 1990.

Weddell, James. *A Voyage Towards the South Pole*. Newton Abbot, Devon, England: David & Charles, 1970.

Chapter 4: So Remorseless a Havoc

Borchgrevink, Carsten. *First on the Antarctic Continent*. Montreal: McGill-Queen's University Press, 1980.

Day, David. *The Whale Wars*. London: Routledge & Kegan Paul, 1987.

Ellis, Richard. *Men and Whales*. New York: Knopf, 1991.

McNally, Robert L. *So Remorseless a Havoc: Of Dolphins, Whales, and Men*. Boston: Little, Brown, 1981.

Murdoch, W. G. Burn. *From Edinburgh to the Antarctic: An Artist's Notes and Sketches During the Dundee Antarctic Expedition, 1892–1893*. Bungay, Suffolk, England: Paradigm Press and Bluntisham Books, 1984.

Neider, Charles. *Edge of the World: Ross Island, Antarctica*. New York: Doubleday, 1974.

Nordenskjöld, Otto. *Antarctica: Or Two Years Amongst the Ice of the South Pole*. London: C. Hurst, 1977.

Payne, Roger. *Among Whales*. New York: Scribner, 1995.

Reader's Digest. *Antarctica: The Extraordinary History of Man's Conquest of the Frozen Continent*. Surry Hills: Reader's Digest, 1990.

Rosove, Michael H. *Let Heroes Speak: Antarctic Explorers, 1772–1922*. Annapolis, Md.: Naval Institute Press, 2000.

Ross, James Clark. *A Voyage of Discovery and Research in the Southern and Antarctic Regions During the Years 1839–1843*. 2 vols. Newton Abbot, Devon, England: David & Charles, 1969.

Ross, M. J. *Ross in the Antarctic*. Whitby, North Yorkshire, England: Caedmon of Whitby, 1982.

Small, George L. *The Blue Whale*. New York: Columbia University Press, 1971.

Tønnessen, J. N., and A. O. Johnsen. *The History of Modern Whaling*. London: C. Hurst, 1982.

Chapter 5: The Last Wilderness

Bertrand, Kenneth J. *Americans in Antarctica, 1775–1948*. New York: American Geographical Society, 1971.

Brown, Paul. *The Last Wilderness: Eighty Days in Antarctica*. London: Hutchinson, 1991.

Dodds, Klaus. *Geopolitics in Antarctica: Views from the Southern Oceanic Rim*. Chichester, West Sussex, England: Wiley, 1997.

Dufek, George J. *Operation Deepfreeze*. New York: Harcourt, Brace, 1957.

Fox, Robert. *Antarctica and the South Atlantic: Discovery, Development, and Dispute*. London: British Broadcasting Corporation, 1985.

Freedman, Lawrence, and Virginia Gamba-Stonehouse. *Signals of War: The Falklands Conflict of 1982*. Princeton, N.J.: Princeton University Press, 1991.

Fuchs, Vivian. *Of Ice and Men: The Story of the British Antarctic Survey, 1943–1973.* Oswestry, Shropshire, England: Anthony Nelson, 1982.

Hastings, Max, and Simon Jenkins. *The Battle for the Falklands.* New York: Norton, 1983.

Joyner, Christopher C. *Governing the Frozen Commons: The Antarctic Regime and Environmental Protection.* Columbia: University of South Carolina Press, 1998.

Joyner, Christopher C., and Ethel R. Theis. *Eagle over the Ice: The U.S. in the Antarctic.* Hanover, N.H.: University Press of New England, 1997.

Knight, Stephen. *Icebound: The Greenpeace Expedition to Antarctica.* Auckland, New Zealand: Century Hutchinson, 1988.

May, John. *The Greenpeace Book of Antarctica: A New View of the Seventh Continent.* London: Dorling Kindersley, 1988.

Peterson, M. J. *Managing the Frozen South: The Creation and Evolution of the Antarctic Treaty System.* Berkeley: University of California Press, 1988.

Quartermain, L. B. *New Zealand and the Antarctic.* Wellington, New Zealand: A. R. Shearer, 1971.

Quigg, Philip W. *A Pole Apart: The Emerging Issue of Antarctica.* New York: New Press, 1983.

Shapley, Deborah. *The Seventh Continent: Antarctica in a Resource Age.* Washington, D.C.: Resources for the Future, 1985.

Siple, Paul. *Ninety Degrees South: The Story of the American South Pole Conquest.* New York: Putnam, 1959.

Splettstoesser, John F., and Gisela A. M. Dreschhoff, eds. *Mineral Resources Potential of Antarctica.* Washington, D.C.: American Geophysical Union, 1990.

Stokke, Olav Schram, and Davor Vidas, eds. *Governing the Antarctic: The Effectiveness and Legitimacy of the Antarctic Treaty System.* Cambridge, England: Cambridge University Press, 1996.

Suter, Keith. *Antarctica: Private Property or Public Heritage?* London: Zed Books, 1991.

Chapter 6: Crude Awakening

Chasan, Daniel Jack. *Klondike '70: The Alaskan Oil Boom.* New York: Praeger, 1971.

Coates, Peter A. *The Trans-Alaska Pipeline Controversy: Technology, Conservation, and the Frontier.* Fairbanks: University of Alaska Press, 1993.

Cole, Dermot. *Amazing Pipeline Stories: How Building the Trans-Alaska Pipeline Transformed America's Last Frontier.* Seattle: Epicenter Press, 1997.

Cook, James. *The Journals of Captain James Cook on His Voyages of Discovery.* 4 vols. Edited by J. C. Beaglehole. Woodbridge, Suffolk, England: Boydell & Brewer, 1999.

Davidson, Art. *In the Wake of the Exxon Valdez: The Devastating Impact of the Alaska Oil Spill.* San Francisco: Sierra Club Books, 1990.

Dmytryshyn, Basil, and E. A. P. Crownhart-Vaughan. *The End of Russian America: Captain P. N. Golovin's Last Report, 1862.* Portland: Oregon Historical Society, 1979.

Ford, Corey. *Where the Sea Breaks Its Back: The Epic Story of Early Naturalist Georg Steller and the Russian Exploration of Alaska*. Seattle: Alaska Northwest Books, 1996.

Huntford, Roland. *The Last Place on Earth*. New York: Atheneum, 1985.

Keeble, John. *Out of the Channel: The Exxon Valdez Oil Spill in Prince William Sound*. New York: HarperCollins, 1991.

Lethcoe, Jim, and Nancy Lethcoe. *A History of Prince William Sound, Alaska*. Valdez, Alaska: Prince William Sound Books, 1994.

Loughlin, Thomas R., ed. *Marine Mammals and the Exxon Valdez*. San Diego: Academic Press, 1994.

McCracken, Harold. *Hunters of the Stormy Sea: The Violent History of the Sea Otter Hunters of Alaska*. New York: Doubleday, 1957.

Miller, Debbie. *Midnight Wilderness: Journeys in Alaska's Arctic National Wildlife Refuge*. Seattle: Alaska Northwest Books, 2000.

Mirsky, Jeannette. *To the Arctic! The Story of Northern Exploration from Earliest Times*. Chicago: University of Chicago Press, 1970.

Naske, Claus-M., and Herman E. Slotnick. *Alaska: A History of the Forty-Ninth State*. Norman: University of Oklahoma Press, 1987.

Picou, J. Steven, Duane A. Gill, and Maurie J. Cohen, eds. *The Exxon Valdez Disaster: Readings on a Modern Social Problem*. Dubuque, Iowa: Kendall/Hunt, 1997.

Roderick, Jack. *Crude Dreams: A Personal History of Oil and Politics in Alaska*. Seattle: Epicenter Press, 1997.

Rogers, George W., ed. *Change in Alaska: People, Petroleum, and Politics*. Fairbanks: University of Alaska Press, 1970.

Savours, Ann. *The Search for the Northwest Passage*. New York: St. Martin's Press, 1999.

Strohmeyer, John. *Extreme Conditions: Big Oil and the Transformation of Alaska*. New York: Simon & Schuster, 1993.

Yergin, Daniel. *The Prize: The Epic Quest for Oil, Money, and Power*. New York: Touchstone, 1993.

Chapter 7: The Ends of the Earth

Arctic Monitoring and Assessment Programme, *Arctic Pollution Issues: A State of the Arctic Environment Report*. Oslo: Arctic Monitoring and Assessment Programme, 1997.

Benedick, Richard Elliott. *Ozone Diplomacy: New Directions in Safeguarding the Planet*. Cambridge, Mass.: Harvard University Press, 1998.

Colborn, Theo, Dianne Dumanoski, and John Peterson Myers. *Our Stolen Future: Are We Threatening Our Fertility, Intelligence, and Survival? A Scientific Detective Story*. New York: Dutton, 1996.

Gibson, Margie A., and Sallie B. Schullinger. *Answers from the Ice Edge: The Consequences of*

Climate Change on Life in the Bering and Chukchi Seas. Anchorage: Greenpeace and Arctic Network, 1998.

Gribbin, John. *The Hole in the Sky: Man's Threat to the Ozone Layer.* New York: Bantam Books, 1993.

Gribbin, John. *Hothouse Earth: The Greenhouse Effect and Gaia.* London: Bantam Press, 1990.

Hall, Colin Michael, and Margaret E. Johnston, eds. *Polar Tourism: Tourism in the Arctic and Antarctic Regions.* Chichester, West Sussex, England: Wiley, 1995.

Harris, Colin, and Bernard Stonehouse, eds. *Antarctica and Global Climatic Change.* London: Belhaven Press, 1991.

Hickson, Ken. *Flight 901 to Erebus.* Christchurch, New Zealand: Whitcoulls, 1980.

Houghton, John. *Global Warming: The Complete Briefing.* Cambridge, England: Cambridge University Press, 1997.

Lindblad, Lars-Eric, and John G. Fuller. *Passport to Anywhere: The Story of Lars-Eric Lindblad.* New York: Times Books, 1983.

Mahon, Peter. *Verdict on Erebus.* Auckland, New Zealand: Collins, 1984.

Naske, Claus-M., and Herman E. Slotnick. *Alaska: A History of the Forty-Ninth State.* Norman: University of Oklahoma Press, 1987.

Nilsson, Annika. *Ultraviolet Reflections: Life Under a Thinning Ozone Layer.* Chichester, West Sussex, England: Wiley, 1996.

O'Neill, Dan. *The Firecracker Boys.* New York: St. Martin's Griffin, 1995.

Roan, Sharon L. *Ozone Crisis: The Fifteen-Year Evolution of a Sudden Global Emergency.* New York: Wiley, 1990.

Soper, Tony. *Antarctica: A Guide to the Wildlife.* Old Saybrook, Conn.: Globe Pequot Press, 1996.

Stevens, William K. *The Change in the Weather: People, Weather, and the Science of Climate.* New York: Delacorte Press, 1999.

Weller, Gunter, and Patricia A. Anderson, eds. *Implications of Global Change in Alaska and the Bering Sea Region.* Fairbanks: University of Alaska Fairbanks, Center for Global Change and Arctic System Research, 1998.

Acknowledgments

Many people have contributed to my understanding and appreciation of the polar regions—so many, in fact, that it might be considered churlish to thank some and thus risk slighting others. But nobody ever said I wasn't churlish, so I'll run that risk and—while adding the obligatory caveat that any errors in fact and emphasis contained in this book are mine and mine alone—thank those who have been particularly generous with their time and knowledge, and who have played a special role.

Anne Dingwall started it all. Although I had long been interested in Antarctica, it was Anne, my boss at the time, who set the wheels in motion one rainy night in Amsterdam and paved the way for my first trip to the Southern Ocean. The full details of the unusual—even, on reflection, borderline embarrassing—sequence of events that brought this about are told elsewhere in print, so suffice it to once again acknowledge and thank Anne for the path on which she set me.

Although not directly connected with this book, several people provided great company and insight during my Antarctic voyages. Special thanks among this group to Naoko Funahashi, Bob Graham, Dana Harmon Charron, Athel von Koettlitz, and, above all, Arne Sørensen, with whom I have been privileged to sail to both the Arctic and Antarctic.

Anne Dingwall was also responsible for my first foray to the Arctic, calling me one day while I was visiting a girlfriend in England and advising me that a ship was ready to leave Amsterdam for the waters north of Norway, as soon as I could get to the Netherlands. When I protested (truthfully) that I was sick and couldn't make it in time, Anne countered that the ship could easily put in at Glasgow and pick me up there a couple of days later. That voyage was to the eastern Arctic; six years later, Kalee Kreider and Steve Sawyer conspired to convince me to make my first voyage to the western Arctic—a journey that prompted me to move, almost as soon as it was over, to Alaska.

Melanie Duchin, Dan Ritzman, and Sallie Schullinger not only helped make that expedition an enjoyable one; they also continued to be the best possible friends following my move to Anchorage. I value their friendship more than I can ever say, and Sallie's departure from the Great Land has left a hole in my life that will be hard to fill. In the same vein, thanks also to the many others who help make life in Alaska so enjoyable, not least Paula Huckleberry, Debra McKinney, Pam Miller, Warren Rhodes, Liz Ruskin, Jacqueline Summers, Ken Waldman, and Sam Walton.

A great many people contributed, directly and indirectly, intentionally or otherwise, to my knowledge of the polar regions. Any history, particularly a general one such as this, relies on work done by those before. Those works from which I directly quote are listed in the section titled "Sources." For those many other books on which I leaned heavily, I have attempted, wherever possible, to give adequate acknowledgment in the text. Sometimes, however, the flow of the narrative would have been interrupted by doing so, and in those cases I sincerely hope that their being listed in the "Selected Bibliography" section will prove adequate acknowledgment of the debt I owe them.

I have received an embarrassment of help and information in the course of my research on both polar regions, either through osmosis or through interviews, both specifically for this book and for some articles I wrote previously. These earlier articles informed the writing in some chapters. Thanks, then, to Donny Barber, Institute of Alpine and Arctic Research, University of Colorado; Colin Bull; Sara Callaghan, Sierra Club; Dorothy Childers, Alaska Marine Conservation Council; Robert Childers, Gwich'in Steering Committee; Beth Clark, The Antarctica Project; Harlan Cohen, Office of Ocean Affairs, U.S. Department of State; George Divoky, Alaska SeaLife Center and University of Alaska Fairbanks; David Douglas, U.S. Geological Survey; Melanie Duchin, Lyn Goldsworthy, Dan Ritzman, and Steve Sawyer, Greenpeace; George Durner, U.S. Fish and Wildlife Service; Julie Edlund; William R. Fraser, Montana State University; Marc Gagné; Ray Gambell, International Whaling Commission; Margie Gibson, Arctic Network; Brad Griffith, Glenn Juday, and

Brendan Kelly, University of Alaska Fairbanks; Anne Gunn, Canadian Wildlife Service; Walter J. Hickel, former governor of Alaska; Brenin Humphreys and Samantha Smith, WWF Norway; Marian L. Izzo and Malcolm B. Roberts, Institute of the North; Tom Lohman, North Slope Borough; Frank Kamp; Bruce McKay, SeaWeb; Larry Merculieff; Pamela A. Miller, Arctic Connections; Pamela Kay Miller and Sterling Gologergen, Alaska Community Action on Toxics; Daniel J. Morast, International Wildlife Coalition; Connie Murtagh; John Passacantando, Ozone Action; Bob Randall, Jack Sterne, and Peter van Tuyn, Trustees for Alaska; Liz Ruskin and Ben Spiess, *Anchorage Daily News;* Ted Scambos, National Snow and Ice Data Center, University of Colorado at Boulder; Doug Schneider, Arctic Science Journeys/Alaska Sea Grant; Darrel Schoeling, International Association of Antarctica Tour Operators (IAATO); Sallie Schullinger; Stanley Senner, *Exxon Valdez* Oil Spill Trustee Council; Jeffrey W. Short, National Marine Fisheries Service, Alaska Fisheries Science Center; Betsy Weatherhead, Cooperative Institute for Research in Environmental Sciences, University of Colorado; Kelly Weaverling; Gunter Weller, Center for Global Change and Arctic Systems Research, University of Alaska Fairbanks; Margaret Williams, World Wildlife Fund–US.

Some of the writing on global change issues in the polar regions draws on my previous research and reporting for the Discovery Channel Online, at http://www.discovery.com, and for *BBC Wildlife, E Magazine, Lapis,* and *New Scientist.* Thanks to the editors at those outlets, Lori Cuthbert, Roz Kidman-Cox, Jim Motavalli, Ralph White, and Michael Bond and Richard Fifield, respectively, for paying me to write parts of this book long before I realized that was what I was doing.

This book began life as an environmental history of Antarctica. When that project couldn't get off the ground, I considered a similar history of Alaska, and then the two ideas morphed—inspired somehow by a winter evening in Anchorage spent watching HBO and drinking Alaskan Amber—into the book you are now reading. This Damascene revelation occurred just before I made a trip to Washington, D.C., where I mentioned the idea to my friend and colleague Boyce Thorne-Miller. Boyce promptly introduced me to Dan Sayre, editor-in-chief of Island Press, who asked me to submit a proposal. Many thanks to Boyce for making it happen, and to Dan for showing such immediate enthusiasm for the idea. Dan then passed the baton to Jonathan Cobb, executive editor of Shearwater Books; from the crafting of the proposal through preparation and submission of the manuscript, Jonathan has been an encouraging, enthusiastic, and supportive partner in the process. His suggestions were invariably right on the money, demonstrating his intuitive understanding of the book and its author and contributing to the betterment of both. Pat Harris proved an alert fact checker and sensitive copyeditor, and I am extremely grateful to her for giving the manuscript its final touches.

Finally, none of this would have been possible without the support of Sea-Web, for which I edit a monthly newsletter, *Ocean Update*. As well as being a rewarding undertaking in itself, my involvement with SeaWeb gives me the financial freedom to take on projects such as this book. My ongoing gratitude to Tom Johnson, who makes sure the paychecks keep coming my way, and to Vikki Spruill, for annually and unquestioningly signing my contract renewal letter.

Kieran Mulvaney
Anchorage, Alaska

Index